Secret Agents Past

The Parting Of The Waters

Ken Stange

BOOKS BY

KEN STANGE

A Smoother Pebble, A Prettier Shell (Penumbra Press)

Advice To Travellers (Penumbra Press)

Bourgeois Pleasures (Quarry Press)

Bushed (York Publishing)

Cold Pigging Poetics (York Publishing)

Colonization Of a Cold Planet (Two Cultures Press)

Embracing The Moon: 25 Little Worlds (Two Cultures Press)

Explaining Canada: A Primer For Yanks (Two Cultures Press)

God When He's Drunk (Two Cultures Press)

Going Home: Cycling Through The Heart Of America (Two Cultures Press)

Love Is A Grave (Nebula Press)

More Than Ample (Two Cultures Press)

Nocturnal Rhythms (Penumbra Press)

The Sad Science Of Love (Two Cultures Press)

These Proses A Problem Or Two (Two Cultures Press)

Secret Agents Past

The Parting Of The Waters

~~

Ken Stange

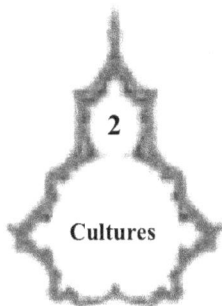

Two Cultures Press

2014

Library and Archives Canada Cataloguing in Publication

Stange, Ken, 1946-, author
       Secret agents past : the parting of the waters / Ken Stange.
-- 1st edition.

Includes bibliographical references.
ISBN 978-0-9939201-0-3 (pbk.)

       1. Creative ability. 2. Art and science. I. Title.

BF408.S726 2014   153.3'5   C2014-906590-6

Acknowledgements

"Redefining Creativity: How Science and Technology Make Us Rethink Creativity" (based on some ideas in this book) was presented and broadcast as a TEDx lecture sponsored by Nipissing University. (2012)

An excerpt from the introduction to this book was published in the anthology of writings on creativity, *Imagination in Action,* edited by Carol Malyon (Mercury Press, Toronto, 2007)

The section "The Solemn Frivolity of Art And Charming Frigidity of Science" was the basis for a presentation in 2010 at the Journal of Arts & Sciences Conference, University of Malta, and subsequently published in the academic journal, *The International Journal of Arts & Sciences.* 3(8), 168 – 174.

Cover Art and Design: Ken Stange.

ISBN: **978-0-9939201-0-3**

This book is dedicated to my not-so-secret agent: Ursula.

Her collaboration in all my creative efforts is a secret all over the block.

Contents

FOREWORD: FOREWARNING

This book is the first in a trilogy on the nature of creativity in the arts and sciences. Organizing my many scattered thoughts on this topic wasn't easy, and I eventually decided to adopt a quasi-chronological approach. So this first book is in some sense about the past, although much still applies. The second book is about the present, which has changed substantially given the increase in resources for the creative. The third and final book consists of speculations on the probable future, even though I know too well that predicting the future is a mug's game.

The three *Secret Agents* books:
> *Secret Agents Past: The Parting Of The Waters*
> *Secret Agents Present: Looking Through A Glass Darkly*
> *Secret Agents Future: Going Where There Be Dragons*

PREFACE: CAVEAT LECTOR

This book is intended for the general reader and not as a scholarly review of 'the literature' on creativity. It is intended as an informal, speculative quasi-philosophical exploration of the nature of creativity in art and in science, spiced up with a few polemics. For this reason, and for the sake of readability, I have kept the customary, often cumbersome, paraphernalia of formal scholarship to a minimum; so the reader will not find any of those intrusive APA format parentheses with the names and dates of research papers. Nor have I banned the use of the first person singular and personal anecdote from my prose, as is required in scientific publications trying to maintain an objective tone. I openly admit to being opinionated and less than coolly objective, for I care passionately about the topic of this book. For the same reason, I've tried to avoid the "Hey, I'm not responsible" passive voice.

Consequently the reader won't find extensive, formally constructed references to support all of the allegedly factual information included in this book. However, most of what I say, not opine, can be easily confirmed independently, given access to The Internet and any decent library. In those cases where some fact or reference is not so easily confirmed by the average reader in our wonderful electronic age, or where I felt it worthwhile to point to interesting material relevant to the topic at hand, I have inserted an adequate bibliographical pointer in a footnote. In most cases I am sure that simply giving the author and the title of a reference these days is sufficient to easily locate full bibliographical information on the Internet.

Thus the policy I have adopted is to use footnotes primarily as a place for extended parenthetical information or comment, including the occasional wisecrack that I couldn't resist. Also, I didn't want to interrupt the flow of the main text with formal bibliographical information, which is why I've also only used the footnotes to give brief author/title pointers to relevant books or articles. Only when directly quoting a writer or when referring to extremely specific material—such as a particular research study—have I felt it necessary to include detailed reference information in the footnote.

Naturally, I accept full responsibility for any errors of fact or interpretation of facts. That I've not called expert witnesses to substantiate all my statements should not be interpreted as my not checking my information. Those who wish to question something I've passed off as factual should find it easy to check up on me, and I'm more than willing to stand corrected. I'm sure there are many experts in the various fields where I've trespassed that will queue up to do just that—and shoot me down. I only ask they don't shoot to kill.

.

So although this book has no references page, I couldn't resist appending recommendations for further reading: I have included a somewhat eccentric, quite eclectic, selected bibliography, sorted by chapter topic. Many of these works have been a source of both factual information and inspiration to speculation.

PROLOGUE: THE MYSTERY PLAY; THE MYSTERY OF PLAY

"Beauty is truth, truth beauty."
—John Keats ("Ode On A Grecian Urn")

"Man is most nearly himself when he achieves the seriousness of a child at play."
—Heraclitus

We are moved to tears. Transported. By certain passages of music, perhaps a Bach concerto wherein sensuously intertwined strands of melody entangle us in their intricate mathematical interstices and make of us glad captives, happily trapped in a contrapuntal web. Or by an old and—more often than not—saccharine pop-tune that triggers vivid memories of our first exploration of the slopes and planes of a young lover's body. By the glistening perfection of an autumnal forest after rain, where the late afternoon light invents a new colour spectrum to display on its canvas of dappled leaves. Or by the sight of a patch of dirty weeds which suddenly brings back the empty lot where as children we imagined ourselves daring, pith-helmeted jungle explorers. Or by the dramatic pathos of a prideful old man orchestrating his own madness by not recognising his daughter, Cornelia, as being his one offspring loving enough to be honest. Or by the silly lines of a bad poem we loved and memorized in grade six before we learned, like King Lear did too late, to distinguish gush from gumption.

We are moved, moved to the highest emotional plane, not so often by life events, by visceral satisfactions or physical suffering, but rather by the contemplation of the world about us, by deeply apprehending both Nature's and our creations: perceptions and memories of perceptions.

It is in creation, recreation, and play—that is where the mystery of human emotion resides. It is from the closing of the eyes that the medieval mystery plays took their name; it is from the opening of the eyes that the play of art and science invoke our deepest emotions. It is reasonable to assume that we share with many other creatures such fundamental emotions as pleasure in being fed when hungry and pain when being physically injured, but it is hard to imagine any other species experiencing far stronger emotions than these *from merely observing*. We are special in that we feel most deeply in contemplation of perceptions and the ideas they invoke, not in the response to the vicissitudes of mundane physical experience. The tears we shed when we get hurt are trivial and more easily forgotten than the tears we shed in response to great art. A speck in our eye may make us cry, but Oedipus blinding himself in tragic despair makes us weep.

.

It is an assumption of this book that this uniquely human experience, which I will call—for lack of a more felicitous term—the "aesthetic experience", lies at the heart of our humanity, is the foundation on which human civilization is built. It is a further assumption that this aesthetic experience is as essential to science as it is to art. These assumptions point to a deep mystery that has always concerned humankind—the mystery of creation. It is most natural that we speculate about from where originates this magic in ideas and images and sounds that so moves us. Whence beauty?

.

One answer is God. In fact, natural beauty and order has been repeatedly put forward as an argument for the existence of God. We all think in terms of cause and effect, even those peoples we condescendingly call primitive or pre-scientific; and consequently anything in nature that is as moving or transporting as a great work of art conjures up an artist. Explain sunsets, demands the religious poet. Enter, stage left, God The Creator, the ultimate artist—or maybe merely one of humankind's impressive creations. But this book is not about natural beauty. It is about the beauty we have created.

.

We know how some of the things that touch our souls came into existence. Human beings made them. We, *Homo sapiens*, created them.

.

It is not the natural phenomenon that most need explaining, for we are already progressing nicely on that project: that is what artists and scientists do for a living. No, what is even more mysterious than the objects in our universe is the awe and wonder these phenomena invoke in us, and, also, very importantly, the mysterious ability of some of us to create things equally able to inspire such awe and wonder. "Only God can make a tree,"* but only Man has made paintings of trees (or developed scientific theories to explain their appearance) that are even more wondrous than most trees. The focus of this book is not on the creations of God (or if you, like me, prefer: Mother Nature or the Laws of Physics). This book is about human creators and human creativity.

.

Descartes' famous self-referential conclusion (as well as initial assumption) was *cogito ergo sum*. This is usually translated as "I think; therefore I am." But an equally valid translation is "I imagine; therefore I am." Or try *creo ergo sum*: I create therefore I am. It is our ability to create that makes us imagine ourselves God-like. It is what distinguishes us from the rest of God's creatures. We *imagine* ourselves. We *imagine* ourselves unique.

.

.

Another book on creativity? Pseudo-scientific books on creativity almost rival diet and cook books in popularity. 'Creative' has become a buzz word littering the titles of every bookstore's Self-Help Section: *How To Tap Your Creative Potential; Creative Problem Solving; Creative Deck Construction; Creative Divorce.* (It would not surprise me to come across a book entitled *Creative Death and Dying.*) Some writers (e.g., Osborn, De Bono†) and their disciples have made astoundingly successful careers purportedly teaching people how to be creative.

.

Educationists prattle on and on about how everybody is creative, while ironically disproving their own contention with the cliché-crippled style they use to preach this sermon.

.

Psychologists write books and papers claiming to explain the nature of creativity, but like several astonishingly humourless 'experts' on humour that I've had the misfortune of hearing speak, most social scientists writing on creativity are so obviously lacking in the characteristic they have chosen to study that one is inevitably reminded of the proverbial shabbily-dressed tailor, of the shoemaker whose soles flap flap as he walks.

.

Creativity. Debased as the word may have become, it remains the best word to describe what is responsible for humankind's noblest accomplishment: civilization.

---

* Or at least according to Joyce Kilmer's 1914 poem "Trees"!
† Both these 'teachers' of creativity will be discussed later.

But then what defines a civilization? Certainly not the conquest or occupation of territory, for by such definition the Huns and Visigoths, for that matter insects or viruses, would be of greater significance than the Ancient Greeks. And for my purposes, I will not use the word in the technical sense that some historians and archaeologists do: a stable society of some longevity with cities of a certain minimum size.

By civilization I mean a commitment to understanding and to morality. The former is prerequisite to the latter, and it is the former that is dependent on creativity. *Civilization is made possible by understanding, and there are two roads to understanding: art and science.* These are the things that matter. Humankind creates what Nature hasn't and is moved by it. This is the mystery that I hope to address in this book—and that mystery goes by the name of creativity.

It is important to make clear immediately that I am not forgetting about or excluding philosophy, mathematics, or the so-called 'humanities' when referring to art and science as the definitive characteristics of civilization. These other noble endeavours straddle the two domains of art and science, and their practitioners are not exclusive residents of one or the other. Philosophy, for example, is the ultimate human commitment to understanding, and some might even argue that it subsumes both art and science. Mathematics I will tend to treat as a science, although it differs radically from science in being largely non-empirical and has more in common with the methods of philosophy than those of scientific investigation. What about the other humanities? They are primarily arts, albeit occasionally scientific in their application, so I certainly am not implying they are irrelevant in any definition of civilization.

When I say art and science ultimately define civilization, I mean only that the greatest achievements of our species are to be found in the successful search for understanding and apprehending truth, beauty, and the good, moral life, and that the two primary ways of conducting this search are art and science. Creativity is the characteristic we attribute to the best of these searchers, but artistic creativity and scientific creativity aren't identical. This book is about how this attribute manifests itself in artists and scientists.

My approach is eclectic—and I hope synthetic—drawing on traditional philosophy, history, biography, all the arts—and sciences.

These all seem relevant sources of insight, perhaps because of my own background.

I am a poet and writer. (I'm also, to a far less competent and committed extent, a visual artist and a social scientist.) But I find admitting to the vocation of poet oddly embarrassing. At least one of my teaching colleagues and friends, who proudly and unabashedly refers to himself as a scientist, always introduces me as "a poet" in a tone that suggests it's some kind of joke, a pretentious claim. I am painfully aware of the presumption of saying: "I am a poet". It is to say: "I am a Creator."

I am not a religious man (quite the contrary), so my discomfort isn't from a sense of sacrilege. Nor is it due to modesty, for, as my friends frequently inform me, I'm not particularly humble. It is just that I am made uncomfortable by the current stereotype (unfortunately somewhat valid) of poets, or most artists, as pompous, selfish, and self-important. My own position is that I just 'happen' or choose to make poems. That I can say without embarrassment. I construct my poems and fictions and essays the same way a homebuilder constructs a house. It involves some skill, many helpers, and some vision. Call it creativity. I believe it sometimes results in something of value: the invoking of an aesthetic experience. But I am proud of my work, not of my 'creativity', not of being a Creator, certainly not of my mimicry of Father God or Mother Nature.

So I must beg the reader's indulgence. To be creative is already presumptuous enough. To try to be creative about creativity is infinitely so. It is, as well, a risky business, for like what computer programmers call recursion, like what mathematicians call infinite regression, like the famous Quaker Oats box with a picture of itself on the cover, like a mirror reflected in a mirror, any attempt to creatively explore creativity entails wandering in the Land of Paradox. Creativity is riddled with riddles.

One major theme of the *Secret Agents* is that the presumption, the *hubris* that seems essential for the creation of works of art and science *should* be tempered by an awareness of the other agents, usually secret agents, affecting the creation. No human creator works alone or with a totally free hand. This book isn't intended as a peon to creative people, for what is really important is what they have produced, not them as individuals. They were the fortunate offspring of genetic and environmental factors that made possible what they accomplished.

(And, if you believe in it, the exercise of their free will.) Creation may very well be humanity's unique talent, but no artist or scientist works alone and so excessive personal pride is unjustified.

.

The reader will find many reflections on, appreciations of, and even pointless, petty complaints about the non-human collaborators with which all artists and scientists must work. "The Secret Agents" of the title refer to our largely unacknowledged collaborators—and censors. They are the ghosts in the human machine. They are the mathematical formulae we discover when we scratch through the surface of mundane reality. They are the strange, not commonsensical, but immutable Laws of Nature. They are the limits and inveterates of our perceptual apparatus. They are both the inherited and the conditioned pathways (or ruts) in our brain's neural networks. They are the ghostwriters behind the creation of our best literary works. They are the sensorimotor circuits that guide the hand that guides the brush as it sweeps the canvas. They are the cosmic composers (who obviously studied calculus) of the music of the spheres. They are the biological imperatives that silently legislate our selection of themes.

.

The two most important of these largely unacknowledged creative collaborators are science and art themselves, and this is the major theme of this work: the aid to artists provided by science and the assistance given to scientists by art. The "Two Cultures" of art and science are both infiltrated by secret agents from the other side.

.

The future of civilization may well be determined by how we interact with these, our silent partners, usually unremarked and unappreciated agents, involved in the creative process. In some cases we must overcome their prejudices, while in other cases we must subordinate our own childish preferences to their superior wisdom. In all cases we should acknowledge their contribution.

.

In the beginning there was light. Lately we have been blocking that light with an overblown image of ourselves. The egotism of contemporary artists is notorious, and scientists, though less flamboyant, are almost as self-centred. One might say that the Renaissance 'put things in perspective', as did the scientific revolutions that followed. (Copernicus and then Darwin certainly put in perspective man's position in the grand scheme of things.) The time is ripe to once again realign our perspective. As the poet

Leonard Cohen says "There is a crack in everything. That's how the light gets in."

Creativity has always been viewed as humankind's most powerful tool, what has given us dominion over the earth. But 'tool' may be a euphemism: creativity is our *weapon*—against barbarism and annihilation. However, it is becoming more and more apparent that it is a double-edged sword. In the arts this is evidenced by the retreat from beauty into parody, irony, and egotism. In science the advances in biotechnology and applied physics are so alarming that quite reasonable men and women are suggesting what seems unthinkable: we consider curbing certain creative endeavours because they are just too dangerous—for our species, even for our planet.

Lawrence Ferlinghetti has a wonderful poem, simply titled "11" in his book *Coney Island of the Mind* that begins: *"The world is a beautiful place to be born into..."* And it is. Partially because of creations for which human beings can claim no credit, but also partially because of human creations—including Ferlinghetti's poem. But I mustn't quote out of context: the poem continues

> *if you don't mind happiness*
> *not always being*
> *so very much fun*
> *if you don't mind a touch of hell*
> *now and then*
> *just when everything is fine*
> *because even in heaven*
> *they don't sing*
> *all the time*

I have divided *Secret Agents* into three books: *Secret Agents Past*; *Secret Agents Present*; and *Secret Agents Future*. Metaphorically they are three panels (if a painting) or panes (if a window) of a triptych such as adorned medieval cathedrals. The first panel portrays the historic record, including that fall from grace, the division of science and art that occurred sometime in the not too distant past. The second, central panel is a picture of the current world of creativity, like a busy painting by a modern Brueghel—or perhaps Bosch—peopled with the secret agents that are helping or hindering, but always shaping, the activities of artists and scientists. The final panel is a roughly drawn, somewhat fantastical, map with what I believe to be the most interesting, albeit often dangerous, of uncharted lands, highlighted for future exploration.

THE PARTING OF THE WATERS
(THE DIVISION OF ART AND SCIENCE)

THE PAST. *The first pane in the triptych: a critical and sometimes painful examination of creativity in the arts and creativity in the sciences, including the reasons for their unfortunate separation and the possibilities of a meaningful reconciliation—despite the siren temptations to adultery and idolatry from such pretenders to creative androgyny as the practitioners of the quasi-science of psychology.*

THE PARTING OF THE WATERS

"Water, water, everywhere,
Nor any drop to drink."
—Samuel Coleridge (*The Ancient Mariner*)

"Between the idea
And the reality
Between the motion
And the act
Falls the Shadow"
—T. S. Eliot (*Four Quartets*)

Thales, often considered the first Greek philosopher, postulated that the world was made of water. At first encounter this seems a strange and naive idea; but if water is compared to the three other ancient elements, there is a certain compelling logic to its selection as the ultimate element. Fire is a rare thing and rarefied. Air just does not seem substantial enough to be considered the ultimate building block of all material existence. And earth, while certainly substantial, probably did not appear as limitless as the sea must have seemed to this Milesian philosopher as he contemplated a sunset over the Aegean. Indeed we now know what Thales might have suspected: two-thirds of our planet's surface is covered by water. So Earth is very little earth: its surface far more water than land. Actually it is the thin skin of 'earth' on a mere third of our large wet rock that is the anomaly, an anomaly that only surfaced relatively recently in the history of our planet.

Yes: water, water, everywhere. Deep water. And much of it undrinkable. Not because it is salt water. But because it is not the water used to quench mundane thirst. It is the water of appreciation and explanation: the water of art and the water of science.

WHAT IS WATER?

Water is many things but at least three clearly distinct entities.

There is the water of art. It is the doldrums water of the Sargasso Sea in Coleridge's *Ancient Mariner* and it is the magical, watery highway of Kenneth Graham's *Wind In The Willows*. It is the churning, oily waters in a Turner seascape and it is the placid Venetian canal waters in a Tintoretto. It is the melodic, aural water of Debussy's haunting tone poem, *La Mer,* and it is the bubbling brook water of Schubert's *Trout Quintet*. It is the water of literature and painting and music. It is water that is not water, composed of substances fluid in a different way: words, paint and musical notes. But nevertheless these waters are deep, intensely real, and capable of quenching a very different kind of thirst.

And then there is the water of science: a tiny molecule composed of two gases! (Ironically Thales' 'ultimate' element is actually made of 'air'!) Water has two parts hydrogen and one part oxygen. As if this is not bizarre enough, science informs us that these elemental gases are themselves composed of unimaginably tiny electrons, neutrons and protons, and furthermore and further down even these subatomic particles are really no more than probability waves in the Wonderland World of quantum physics. Water as abstraction, as mathematical formulae, but nevertheless deeply, intensely real. Just as the thirst for knowledge is real.

And finally there is sensate water: the water of our experience, from womb to suburban swimming pool. The water we actually drink when thirst isn't just a metaphor.

So which of these waters is *most* real? The magical, metaphoric water of art? Or the abstract water of science? Or the water of direct sensory experience?

Probably the commonest response would be that what we perceive, through our complex perceptual apparatus, is the ultimate reality. Most of us, no matter how hard-headed and rational we fancy ourselves, credit our direct sensory experience more than anything else. Everything else, it is tempting to think, can be relegated to a mere attempt at explanation (if science) or at simulation (if art).

SENSATE WATER

It is a brutally hot summer day. Surely the feel of silky coolness on your sunburned skin as you dive into a northern lake or the crisp taste of spring water on your parched palate or the sight of whitecaps glistening in the sun is the definitive *reality* of water!

.

Of course I just stacked the cards in that last paragraph by trying to inspire recollections of *intense* experiences of water. Now consider instead those many tepid times when water was nothing. For example, you are washing dishes while engaged in a heated discussion with your spouse. Or you are sipping a glass of water while engrossed in a book. If you compare your experience of water at times such as these, times when water was something scarcely noticed because your attention was elsewhere, then sensate water is something far, far less experientially intense than what you felt the first time you read *The Ancient Mariner* or *Moby Dick* or *Huckleberry Finn* or *Jaws*.

.

Consider, too, what physiologists and psychologists say is *really* happening when you perceive water, the cold hard facts about what is 'actually' happening when on a summer day you dip your hand in a cool brook. Neural receptors located in your dermis are triggered by the temperature *change*. This results in a flow of sodium ions through the membrane of an afferent (or sensory) neuron, which in turn causes a microvoltage differential to travel along the membrane of this nerve cell. When this charge arrives at the axon terminal of the cell it stimulates cellular vesicles (little bubbles full of organic chemicals) to dump a stew of neuro-transmitters into the synaptic space. If enough of these neuro-transmitters (arriving from different afferent neurons with receptors in the same vicinity of the dermis) happen to be splashing around in a synaptic space by the dendritic end of another connector neuron, the whole process will be repeated—and the next neuron in the chain will carry the 'message' along. If all goes well the process will continue up your spine to your brain where that midbrain switching station called the thalamus will route the information up to the association neurons of the particular area of the sensory cortex that is dedicated to how you feel. (It is reassuring to know something cares about how you feel!) Then somehow, miraculously the stimulation of these neurons, the actions of the tiny sodium ions leaking through the cell membranes, generates the *experience* of coolness on your skin. Neuroscientists refer to this experience by the technical, but poetic, term *appreciation*.

So how dare we anthropocentrically claim that our particular experience of water, an experience that is demonstrably different from that of an amoeba or paramecium (who have no neurons) or that of electric eels or whales (who have specialized receptors capable of detecting attributes of water not even dreamt of in our philosophy), is the ultimate reality of water?! Even without having to resort to sobering science, epistemologists have repeatedly warned of the danger of confusing perception with underlying reality, what Immanuel Kant called, respectively, phenomenon and noumenon.

So let us resign ourselves to the fact that Water is not the specific water of our sensory experience. The hardwired circuitry of our nervous system creates the water of sensate experience. It is not ultimate Water. So where then do we look for *real* Water?

SCIENTIFIC WATER

Is the water of science, H_2O, more real? Since it is science that explains what is *really* happening (or at least so most educated people in industrialized nations believe) when you feel, taste or see water, does it not make sense to look to scientists for the truth about water—instead of naively assuming our quirky, neurologically-determined and anthropocentric 'experience' of water is what water is *really* about. The water of modern science may be abstract, even arcane, but scientific abstractions which we may not understand have significant effects which we observe daily, everything from television to medicine.

.

Those who are not scientists (or not very good scientists) put a great deal of faith in science as the only true path to reality. But those who actually do science at the cutting edge (where it bleeds into philosophy) are far more circumspect. Consider physicists.

.

The overwhelming majority of contemporary physicists endorse what is called the Copenhagen Interpretation,* a dictum usually attributed to Neils Bohr, Nobel laureate, the originator of modern atomic theory, and founder of The Copenhagen Institute for advanced studies in physics. This dictum is bipartite.

.

The first part of the Copenhagen Interpretation is the startling claim that *there is no deep reality*, that all there *really is*—is appearance.† What the philosopher Immanuel Kant called the *ding an sich* (literally "thing-in-itself") just plain does not exist. To use Kant's terms, there are no noumena (underlying realities); there are only phenomena (observations). Plato maintained that what we saw and thought to be reality was only shadows on the wall of a cave. Modern physicists would not only agree, but would go much, much further: they would argue that *there are no objects with distinct shapes casting those shadows*. We draw the shadows ourselves by the act of looking at them.

.

The second part of the Copenhagen Interpretation seems even more outlandish. It states that *reality is created by observation*. Nothing, or more accurately no dynamic attribute of anything, exists in the

---

* There are some notable exceptions, including Einstein.
† My wording and explanation of the Copenhagen Interpretation is largely based on Nick Herbert's *Quantum Reality: Beyond The New Physics*. However, his description matches well with that of numerous other explicators of quantum mechanics.

absence of observation.* Some physicists would even insist that a conscious, sentient being is prerequisite to the existence of anything. At first this sounds like crazy talk at the Mad Hatter's tea party, but there are some very eminent, very sober people at this party. In fact, John von Neumann, often said to be one of the clearest thinkers and mathematicians of the twentieth century,† the author of what is called the Bible of Quantum Mechanics (*Die Mathematische Grundlagen der Quantenmechanik*), is usually credited with originating the idea that "physical objects would have no attributes...if a conscious observer were not watching them."‡

Imagine a lake, hidden deep in a virgin forest of The North. Imagine the attributes of that lake: isolation, water as pure as polar ice, blue as cobalt. Sorry, the most scientific of scientists would say, but such a lake *does not exist* until a human being, seeking after ultimate wilderness, reaches the light at the end of a long portage and gazes upon it. And then, of course, it loses that first mentioned attribute: it is no longer truly isolated. Or, to change the metaphor: for a physicist, there can be no virgins until they have been deflowered.

---

* Physicists do not go quite as far as the philosopher Bishop Berkeley: they do not deny the ultimate existence of, for example, electrons, but they do maintain that electrons do not have, for example, any <u>real precise</u> position in space until they are observed.

† John Allen Paulos in *Beyond Numeracy* goes even further: "referred to by some as the smartest person who ever lived." Von Neumann will be discussed in some detail in a later chapter.

‡ This is Nick Herbert's paraphrase (in his *Quantum Reality*) of Von Neumann's position. Whether the critically important observer really need be a sentient being is a philosophical quagmire.

ARTISTIC WATER

So if science, by its own admission, fails to serve up objective reality, then where oh where can we turn—but to art? Undoubtedly there is a special understanding to be had in art. I have seen the sea; have swum in it, even canoed on it. I have read scientific works dealing with the sea. But, personally, my strongest impression—and what feels like my deepest *understanding* and profoundest experience of it—comes from Melville and Coleridge and Conrad.
.

I think that to view art as imitation or simulation is anachronistic, naive. I will address this issue later, but for now assume that art is not imitation but creation, that Conrad's sea (or Turner's or Debussy's) is not a representation or explanation of water, but rather a reinvention of it. If in physics, the nature of water does not exist until observation creates it, in art the nature of water does not exist until the artist creates it.

ARE ALL OASES MIRAGES?

So where does this leave us? Water, water, everywhere, and not a drop that's real. Yet the Grand Canyon is there—because (or so we all believe) the rushing waters of the Colorado River have, over eons, *created* it.

And yet water, that original element, that daily experience, that component of our body that accounts for 65% of each of us—is not something real.* It is created by us, by our bodies or our minds. And it is created again and again, each and every time in profoundly different guises.

To review the options:
- There is the water of sensation or experience: our bodies, our nervous system create this water of subjective experience.
- There is the water of science: scientists create formal, abstract structures from numbers and symbols and logical relationships they claim define water.
- There is the water of art: artists create a water from unwatery things (words, paint, musical notes) capable of seeming deeper than the deepest pools of direct experience.

The important thing to note is that in all of these cases there is creation. This is not to maintain, in pure Idealist fashion, that there is absolutely no external reality. But it *is* to maintain that the *attributes* of this reality are created by us.†

In the first case (the most immediate, sensate reality) what is created is created not by us but by our bodies: the creation of reality by neural circuits does not contribute to the collective human enterprise.

---

* So maybe we aren't (as described by the universal translator in a "Star Trek" episode) merely "ugly giant bags of mostly water."

† As already remarked, few physicists would agree with the first and most radical Idealist, Bishop Berkeley, that there is no external reality at all, only mind. What the majority of physicists will say, however, is that dynamic attributes (variables) do not exist until observed: an electron, all electrons, have a given, real atomic weight (the same for all), but no electron has a particular, distinguishing position or momentum until the moment of observation. Similarly, the artist does not invent the sea, but a particular sea, and what you feel or don't feel when you dive into a lake depends on your individual physiology.

Although some hedonists have tried to raise sensuality to the level of an art or science, no one receives a Nobel Prize for feeling good or intensely.

.

It is only in the other two cases that the thing created is not merely subjective and thus trivial. For any person whose teleology is not based on God, but does believe that human life has an ultimate purpose, that ultimate *raison d'etre* must be the invention of the world through art and science. Why then do these two noble human endeavours so often seem to be in conflict?

WHEN THE WATERS OF ART AND SCIENCE PARTED

In May of 1959, C. P. Snow, novelist and scientist, gave The Rede Lecture at Cambridge. As was traditional, the following day the lecture was published in paperback. For several months there was little response, the usual fate of such lectures. Then gradually at first, but soon exponentially, a wave of reaction, often extremely hostile, rose in the intellectual community of England, then The Continent and The United States. Snow describes his surprise and discomfort: "By the end of the first year I began to feel uncomfortably like the sorcerer's apprentice. Articles, references, letters, blame, praise, were floating in—often from countries where I was otherwise unknown." He goes on to observe how "a nerve had been touched almost simultaneously in different intellectual societies, in different parts of the world."*

What was this exposed nerve he'd touched that could produce such a tidal wave of response? The name of the lecture supplies the answer: *The Two Cultures*. The phrase has become a part of the standard intellectual lexicon. The two cultures are, of course, science and art. C. P. Snow had the temerity to point out a nasty little fact that many knew, but no one had so baldly stated in public or in print: scientists and artists had parted ways, had split into two cultures, and were profoundly ignorant of—and often extremely distrustful of—each other. He had effectively accused all intellectuals of parochialism: the deepest cut of all.

If he had accused scientists of being illiterate and uncultured, he would have had the whole artistic community backing him up and cheering him. Or if he had accused the artistic community of being ignorant Luddites, he would have had scientists praising his honesty. But, as a member of both cultures, he felt he had the right to criticize both and did so, thus arousing the ire of both.

I first read *The Two Cultures* forty years ago and most of it matched my experience and my perception of the current intellectual climate. I recently reread it, and while I still agree with his initial assessment of the situation, I do believe, as will become evident, that things are

---

* These quotations are from Snow's essay "The Two Cultures: A Second Look" which he published four years after his original lecture.

changing. In fact, I hope this book may make some small contribution to the reconciliation of the two cultures.

The beginning of Snow's essay is a cool, well-reasoned and reasonable assessment of the relationship of the artistic community (what he calls the literary intellectuals) to the scientific community, but this is only Snow's starting point. Three quarters of the essay is an impassioned plea for better, broader education and the humane application of science and technology to alleviate the cruel economic inequalities of the world. That such a fundamentally decent, even commonsensical, piece of writing could enrage so many people is saddening evidence of the extreme egotism and sanctimoniousness of both artists and scientists.

It is an historical question, which I won't attempt to answer here, where and when the rift first occurred. Da Vinci obviously saw no contradiction in being interested in both science and art. But at least as far back as the English Romantic poets, there was frequently expressed hostility between the two camps. For example, Keats certainly was no fan of Newton, accusing him of "unweaving the rainbow."*

One place where I disagree with Snow is when he remarks: "The clashing point of two subjects, two disciplines, two cultures—of two galaxies, so as it goes—ought to produce creative chances. ... It is bizarre how very little of twentieth-century science has been assimilated into twentieth-century art." He is wrong. Modern science *has been* assimilated into, and influenced, modern art—just as art has influenced science. There are secret agents working both sides of the great divide.

---

* Keats is quoted by Haydon in his autobiography as saying of Newton that "...he destroyed the poetry of the rainbow by reducing it to a prism." And in his long poem *Lamia* these lines occur, making it very clear what Keats thought of 'Natural Philosophy' (a contemporary term for science): "Philosophy will clip an Angel's wings, / Conquer all mysteries by rule and line, / Empty the haunted air, and gnomed mine– / Unweave a rainbow..."

THE TWO CREATIVITIES

So this first pane (pun intended) in the triptych that is this book deals specifically with the nature of scientific and artistic creation, but the similarities and differences of creativity in art versus science will be a recurring theme throughout the whole book, because I hope to demonstrate how science is, in fact, a creative collaborator for the artist and vice versa: that science feeds the creative artist and the arts feed the creative scientist. The ongoing nasty spat between artist and scientist is partially based on personality differences, partially on differences in goals, and partially on ignorance and misunderstanding. Only the last can be remedied, but if it is, conflict based on the first two can be ameliorated through an appreciation of how each serves the other.

The simplistic and—like many simplistic explanations—reassuring idea that creativity in art and science are virtually the same is patently nonsensical; instead, there is a complex interaction, a symbiotic relationship, between these two important human enterprises. However, before further considering the schism between art and science, another apparent dualism, one often considered parallel, has to be considered: the relationship of the logical mind to the creative soul, convergent versus divergent thinking, intelligence versus creativity.

And even before that, there is the important question of what precisely I will mean when I use the words "art", "science", and "creativity." It is time for the naming of parts.

THE NAMING OF PARTS

"Today we have naming of parts. Yesterday,
We had daily cleaning. And tomorrow morning,
We shall have what to do after firing…"
—Henry Reed ("Naming of Parts")

"What's in a name? That which we call a rose
By any other name would smell as sweet."
—William Shakespeare (*Romeo And Juliet*)

"It's just a matter of semantics." These words are often invoked in an attempt to end an argument, the reasoning being that if those in dispute simply agreed to the meaning, the definitions, of the words they were using, they would have to come to the same conclusions. Sometimes of course this is true, but these magic words cannot make all disagreement disappear. In fact, the opposite can happen: clarifying one's definition of something can turn apparent agreement into dispute, as when two people say they both believe in God, but further conversation reveals two very different and sometimes hostile conceptions of "God". Radical Muslims and Fundamentalist Christians both believe in "God".

.

Creativity, like God, is a word with many definitions, so it is necessary to examine these many meanings before proceeding. But even before doing this, it is necessary to clarify the definition of "definition" in science.

THE DEFINITIVE OPERATION

One of the key concepts in science is that of the *operational definition*. This refers to the defining of a variable by an action or operation. Consider this imaginary scenario, which is representative of experimental design and the role that definition plays in it.

Let's imagine that three researchers have all heard contradictory folk 'wisdom' about drinking before going to bed: the strait-laced and abstemious maintain that it is not prudent if one wants a good night's sleep; the average person, however, claims that a nightcap is a relaxing and a pleasant cure for mild insomnia. All three researchers decide to test experimentally these two contradictory hypotheses about how alcohol affects "latency to sleep".[*] In an experiment, the suspected cause of something is called the *independent* variable (IV), and the effect it is suspected to have is called the *dependent* variable (DV), because its state is hypothesized to *depend* on the state of the independent variable. In a typical experiment, one can think of the IV as the hypothesized *cause* and the DV as the *result*. In this example, then, the IV is the alcohol and the DV is the latency to sleep. Now before any experiment can be performed these variables have to be operationally defined—"have to" meaning that it is impossible not to do so. The experimenter must manipulate the IV and that manipulation is the operational definition of it. The experimenter must also choose a way of measuring the outcome, the DV, and that method of measurement is its operational definition. Let's assume that all three scientists decide to define the DV as the time it takes from when their subjects lie down in a dark room until they show the EEG pattern associated with Stage 1 sleep.[†]

However, our researchers choose different operational definitions of alcohol. Dr. Whitebread operationally defines alcohol as "one light beer ingested orally fifteen minutes before the subject lies down to sleep." Professor Souse operationally defines the IV as "one pint of Canadian whiskey downed in the hour before the subject lies down to sleep." Mr. Hyde defines the suspected causal agent as "one litre of over-proof rum injected intravenously before the subject is monitored for Stage 1 EEG patterns."

---

[*] Merely saying "how long it takes to go to sleep" just doesn't sound as scientific.
[†] The subjects all have electrodes attached to the scalp that are wired to a polygraph to monitor brain activity.

Let us assume that all three experiments are free of any research design flaws and involve enough randomly sampled subjects to make their conclusions statistically significant at a very high confidence level. The three scientists publish their results in different journals. Dr. Whitebread reports that alcohol has no effect whatsoever on latency to sleep.* Professor Souse reports that alcohol greatly decreases the time it takes to go to sleep: all his experimental group subjects passed out, many before they even finished receiving the experimental treatment.† Mr. Hyde reports that alcohol causes you to go to sleep forever—induces the big sleep.‡

The Temperance Movement finds Mr. Hyde's paper useful to their cause and cites it in their pamphlets on the evil of drink: "Scientists prove alcohol kills!" PR persons for several major distillers slip a few bucks to reporters working for major newspapers, who, mostly being drinkers themselves, gladly pen articles reporting on Dr. Souse's findings, commenting that recent research indicates alcohol is not only good for your heart, it also aids insomnia sufferers. (Dr. Whitebread's findings are reported less frequently than those of his colleagues, because they don't offer support for any particular prejudice.)

Ah and then the editorial writers for the popular press, who are notoriously lacking in imagination and often envious of the attention science gets, gleefully reason from a cursory reading of these reports that scientists simply cannot agree on anything, so should be ignored! Of course, these three scientists are not in disagreement at all: they simply chose to use different definitions for their independent variable. Of the three operational definitions, Dr. Whitebread's probably comes closest to the commonsensical, and his findings thus most relevant to the average person considering a nightcap, but that doesn't make his conclusion the "correct" one.

---

* Of course generally it is difficult to get a study published that doesn't have statistically significant results, i.e., a study that supports the Null Hypothesis.
† The term "experimental treatment" refers to the manipulation of the independent variable, in this case the ingestion of the booze.
‡ Shortly after publication of his results, Mr. Hyde, along with the Ethics Committee that approved his research design, was arrested and charged with Murder In The First Degree. The charge was subsequently reduced through plea-bargaining based on his scientifically sound claim that there was no premeditation: a researcher can't know the outcome of his experiment until it is carried out.

As will become evident, many of the apparently contradictory conclusions of scientific studies of creativity are only explicable once the operational definitions are made explicit. I will try to do so.

SCIENCE AND THE SCIENTIFIC METHOD, MUTT & JEFF

Defining science might seem relatively easy, but any proffered, seemingly reasonable definition, as every philosopher of science knows, is fraught with contradictions and conundrums.

.

The first question to be decided is whether to define science as a *body of facts* (and, if so, what exactly is meant by 'scientific facts') or rather as a *methodology*. A parallel question, to be addressed shortly, exists for art, especially since so many contemporary artists emphasize process over product.

.

I think it most useful and least confusing to consider science as a product; i.e., science is information acquired through the application of the scientific method. When we speak of the arts and sciences, usually we are speaking of accomplishments, specific types of knowledge in the sense I defined it in the opening chapter. However, scientific knowledge is knowledge arrived at by the *scientific method*, and so it matters very much how one defines this process. This book is, after all, about how we create art and science—i.e., about methodology.

.

So what exactly is the scientific method? Its primary, defining, and indisputable characteristic is the reliance on empirical evidence, on observation. This is what endows it with a universality not found in art. However, science is done very differently in the different sciences. It can be very difficult to find features common to the methods of, for example, astronomy and chemistry and psychology, but all scientists gather their data by observation and use observation to test theoretical conclusions. In all cases, the proof is in the taste of the pudding—not in pure logical consistency, as in math, nor in subjective intuition and emotion, as in art.

.

I remember one of my children coming home from school with instructions for doing a science fair project. My first reaction to this formulaic guide (with its "causal question" and "matrix") was annoyance. This was because, knowing something about the diversity of scientific investigation, the over-simplification of the complex enterprise of science offended my pedantic sensibilities. My reaction was silly, of course, and now I'm in the embarrassing position of having to do a similar over-simplification if I am to get done with this business of defining my terms and go on to an examination of the real world of doing science—and art.

So science in a nutshell, here! (Philosophers of science and working scientists, please just grit your teeth or skim over this.)

Jeff, who has a theoretical bent, is walking down the road with his companion Mutt, who has more interest in making observations. They come upon a dead racoon lying on the shoulder of the road. Mutt wants to investigate, sniff around. Jeff prefers to speculate and forms a hypothesis, which is nothing more than a guess as to the relationship between two things *marked with the intention to test this guess*. Hypotheses can come from anywhere—gut feeling or intuition, authority, voices in one's head. Jeff hypothesizes that the raccoon was gunned down—based on his Uncle Ben's authoritative claim that local juvenile delinquents armed with 22 calibre rifles have been shooting birds and other wildlife just to amuse themselves. Formulating a hypothesis is the first step in the scientific method.

The next step is to use deduction to predict something observable that must be true if the hypothesis is true. Bullets leave holes in their victims. If the poor coon was shot, he'd have a bullet hole. What defines a good hypothesis is its empirical testability. Jeff has a good hypothesis, because he can check the carcass for gunshot wounds. But he would've had just as good a hypothesis if he had thought the racoon was dead because a pygmy had shot it with a poison dart. What makes both of these hypotheses good ones is not *provability*; it is their *falsifiability*: either can be shown to be false by examination of the coon corpse. In science one can never prove anything in the sense that a mathematician proves a theorem: one can only find supporting evidence; and one way of finding supporting evidence is to make clearly defined deductions from one's hypothesis that *can be* shown to be false by observation—and then have observation fail to show them false.

And this is just what Mutt does to evaluate Jeff's hypothesis. No bullet wound is discovered, which is a testable and falsifiable prediction from the nasty kid with a gun hypothesis. (No dart wound is found either so the murderous pygmy hypothesis is shown to be false as well.) Jeff, being a good theoretical scientist, then formulates another hypothesis—the road-kill hypothesis. If a passing vehicle killed the coon, then the body would show signs of being struck. The body of the victim indeed does appear to have been struck by something. This does not 'prove' this second hypothesis, but it offers supporting empirical evidence for a falsifiable prediction that was tested and not proven false. So the hypothesis is now elevated to the status of a scientific theory—albeit a weak one. And there it is likely to stay for a long time, unless someone finds a witness to the hit and run—or a local juvenile delinquent admits to bludgeoning the poor creature to death, another hypothesis that would have also been supported by the evidence of battering on the corpse, but is less plausible because a vehicle accidentally striking of an animal on the road is more common than such juvenile viciousness.

.

Scientific method. Formulate a possible (and falsifiable) explanation for an observed phenomenon. Deduce testable (observable) consequences of the hypothesis. Make observations to see if the predicted consequences occurred. If evidence is present, *tentatively* elevate hypothesis to status of theory *and* continue trying to prove it wrong.* If evidence is not present, formulate new hypothesis and repeat procedure.

.

Science. The theories generated by the scientific method.

.

This isn't a philosophical treatise on the nature of science, so these simplistic definitions will have to suffice for now and, I hope make explicit what I mean when I use the terms. The nature of science and doing science, however, are themes that will be examined in more detail later.

---

* In the case of the poor, dead raccoon, probably no one will care enough to subject the road kill theory to any further testing. The history of science is littered with theories of this sort that, when eventually someone takes an interest and does further investigation, are found to be flawed.

BUT IS IT ART?

Defining science and how it is done is difficult enough, but the challenge of finding a working or operational definition for art could drive even an aesthetician to the familiar protesting lament: "I don't know anything about art, but I know what I like." (I suspect that this is the reason contemporary aestheticians so often seem to avoid this question, despite its centrality to their whole discipline.) Why should the meaning of this word present such difficulties—and often incite such vehement debate? And what are we to make of the fact that many cultures don't even have a word for "art"?

Some time ago I wandered through an exhibition of African 'art' in some public museum or gallery in Canada or the United States. Frankly, I can't remember where exactly this was, but I do remember that a central theme of the exhibition was that most of the beautiful objects displayed were produced by anonymous artisans—and in cultures whose language lacked a word for art. Many of the objects were utilitarian or decorative: jewellery, adorned utensils and tools, or statues associated with religious observance. In some cases it was unclear, or at least unexplained, what role the object originally played in its natural environment. Yet these lovely things from cultures where there were no art museums, no conception of art museums, not even a separate word for what we call art, were nevertheless art objects—things possessing the mysterious power to draw the attention and admiration of people from a very different background. The necklaces were not adorning the slender, gracious necks of beautiful young women in the museum: they were mounted on display cases. And I doubt the statuettes were inducing any religious feeling in any of the men in suits or women in jeans that peered at them with unabashed delight. So what was happening here? How is it that these foreign, culturally displaced objects could retain sufficient value so out of context as to induce thousands of people to slap cash or credit card on a ticket desk just for the pleasure of viewing them? What would the people for whom these objects had a value integrated into their lives and view of the world make of this? What would the men and women who created these things make of this phenomenon?

I vaguely remember something in the curator's written commentary, an implied answer to the above questions, which seemed to suggest that Western Civilization's conception of art was artificial, a meaningless conceit. Although I can appreciate the irony (and implicit criticism of both the self-importance of contemporary artists and the idolatry of those things some art critic decides we should call art), I certainly didn't come to the same conclusion. Art is a very useful concept, and that we denizens of the First World do have a word for it is to our credit. The exhibition in fact is confirmation of this.

.

Because a particular language lacks a word for something isn't evidence that the thing doesn't exist. I doubt any of the native languages of the people who produced the work at this exhibition have a word for electron or television. In both cases this is reasonable and understandable, but for very different reasons. They may not have televisions, so of course they won't have a word for it, even if such a thing does exist in my living room. But they do have electrons: they just haven't examined this aspect of their physical world in enough detail to become aware of the need for a word representing such a concept. But lightning still strikes in the heart of Africa, just as art does in the heart of all Humankind.

.

So as with science, it is essential to the forthcoming discussion of creativity that I make my operational definition of 'art' explicit. I do not claim it is the 'correct' definition, only that it is useful, commonsensical, and relevant to a meaningful examination of creativity in the arts. So I will try to show how I arrived at this working definition, but I will not attempt to convince that it is the best or most accurate.

.

One reason so many people blithely but adamantly maintain that a particular thing—for example, an abstract painting by Malevich or Newman—is "not art" is because they personally have no aesthetic response to it. They are not "moved" by it, feel nothing when confronting it. Philosophers use the term "aesthetic experience" to describe this experience of being "moved", to describe the emotional reaction of the viewer or reader or listener to a work of art, to describe what it is people get out of art that draws them to it. Strangely this amorphous thing with the infelicitous label "aesthetic experience" or "aesthetic response" is a positive thing, a thing sought after, even if the label for the emotion elicited is a negative term: it seems reasonable to assume that people don't like to be sad, yet sadness is the term we apply to the aesthetic response evoked by a Greek tragedy or even a tear-jerker movie. Again this suggests a problem with definition: we are obviously using the same word, sadness, for two very different emotional states: one we would like to avoid and another we actively seek out. This is something worthy of further consideration, but for now I want to focus on the importance of the aesthetic experience as a defining characteristic of what we call art.

It seems reasonable to say that only those things that elicit this aesthetic response deserve the label "art". However, this is not to say all things that elicit this response are art: we normally only use the word "art" to describe something created by the hand of Man.* The random play of light on the surface of a pond crowded with lily pads may very well evoke the aesthetic response, a response not, I think, substantially different or easily distinguishable from the response many people have while viewing one of Monet's lily pad paintings.

So my operational and working definition of art is simply anything *created by the human hand* that evokes the aesthetic response. Removing those things that evoke the aesthetic response that are not created by human beings is not to denigrate them; it is simply to try to make the definition match what most people mean by the term—a fundamental principle for the creation of useful operational definitions. I believe it to be a simple, elegant and useful definition; and one most people would have no problem accepting as what we usually mean when we say something is a "work of art". It also has the advantage of avoiding the thorny problem of "beauty"—which confounds any definition of art based on the object instead of the appreciators' response to the object. Furthermore by shifting the emphasis from the cause (the work in question) to the effect (the response to the work), one avoids the confusion caused by the incredible diversity of art forms. The aesthetic responses to a ballet, a painting, a novel, a concerto have much more in common than do the characteristics of the causes of the response. But of course, there are obvious problems associated with my definition.

First, there is the problem exemplified by the man standing in front of Barnett Newman's "Voice of Fire" at the National Gallery of

---

* Perhaps this a good place to say that I will not, for stylistic reasons, be politically correct regarding so-called "gender issues". I find the insistence on the phrases 'him and her' and 'he and she' obtrusive and silly. (I won't even say anything about such grotesqueries as 's/he'.) I'll use the masculine form of the third person singular, and assume the reader will understand that unless the antecedent clause refers to a particular male, the pronoun is intended to refer to both sexes, as it traditionally has. And 'Man", course, means both male and female humans.

Canada and whose only emotional response is anger at the amount of taxpayer's money spent to acquire this work (1.8 million Canadian dollars or well over 2 million USD)—and who would unhesitatingly exclude this painting from the category "art" as I've defined it, because he is most definitely not having an aesthetic experience from viewing the work. It is not difficult to sympathize with this emotional, although not aesthetic, response. The work is a huge, boring painting of a few stripes. And, interestingly, the more one knows about Newman (his other minimalist paintings, his pompous, banal pronouncements), the more one is inclined to share the casual viewer's outrage. But let's return to this fellow in a moment after considering a situation less controversial which might cast some light on this problem.

.

I remember walking out of a movie theatre several years ago after my wife and I had just finished watching the film *Mon Oncle Antoine*. We both spoke simultaneously, I exclaiming how long the movie was, and she expressing, just as vigorously, how unfortunately short it was. In fact the film was fairly typical 'feature length'. My reaction was based on tedium, hers on dismay at the perceived brevity of her intense aesthetic experience. Both our reactions were expressed so strongly we burst into laughter. The point of this anecdote is not that "time flies when you're having fun"; it has to do with our subsequent discussion of the film which gave us both some insight into each other's psyche—as is the nature of such discussions—and that this discussion never deteriorated into a disagreement about whether or not the film "was art". I readily accepted it as a work of art—not because, incidentally, it was considered an 'art film' as opposed to a mere 'movie', but because I had clear and present evidence in my wife's reaction, that even if I found Mon Oncle a boring old coot, he still had enough life in him to elicit this thing called an aesthetic response in my wife.

.

So to return to the fellow in The National Gallery, one can't help but think that his refusal to accept Newman's painting of a few oversized stripes as a work of art might not only be because he is unmoved by it, but also based on incredulity that anyone could have an aesthetic response to it. One could call this The Emperor's New Clothes attitude, the idea that modern art (including literature and music) is a giant hoax perpetrated on those who want to appear sophisticated. It so happens that I too fail to feel anything (except tedium) no matter how long I look at "Voice of Fire", but I am still reluctant to exclude it from the set of things called art, because I have had deep and 'moving' responses to other abstract art, even works more minimalist, that would have left our indignant gallery goer just as cold—and just as willing to deny them the elevated status of Art.

My point is that the default decision for admittance to the set of things called art should be acceptance, not rejection. If we have anyone's word for it that a thing has produced an aesthetic experience, we must credit that thing with being a work of art. In a scientific experiment there are two possible false conclusions: they are called Type I and Type II errors. The former consists of concluding the independent variable has had an effect, when in fact it hasn't; this is sometimes called a "false positive". The latter, sometimes called a "false negative", consists of concluding there is no effect when, in fact, there is one. In science, as in common law, causes are innocent until proven guilty: it is generally considered better "to let ten guilty men (IVs) go free than convict one innocent one".* Scientists generally are most fearful of falsely attributing cause, most worried about Type I errors. In the case of art, however, we should take the opposite stance: we should be more willing to accept a thing as causing an aesthetic response than automatically denying that it could just because it fails to do so for us. Things should be assumed 'guilty' of being art until proven 'innocent'—i.e., aesthetically meaningless to anyone who has apprehended them.

The first obvious objection to my definition is that it is so extremely inclusive as to make the term 'art' virtually meaningless—an objection similar to one I will make about some definitions of creativity. (One is reminded of Andy Warhol's remark that "art is whatever you can get away with.") However, I don't think it as inclusive as it first seems. It is hard to imagine anyone saying they have had an aesthetic response and received pleasure from contemplating most of what they see as they look around them, hard to imagine anyone claiming to being "moved" by a box of staples or a screwdriver or an electrical outlet—to mention some of the things I can see as I type these words.† And other common man-made objects we encounter may well deserve the appellation 'art'—e.g., the

---

* Actually in science the rule of thumb is a ratio of 20 guilty independent variables to 1 innocent one, since the traditional cutoff point for statistical significance is 95% confidence.

† Of course, given 6 billion people on this planet, surely almost any object might trigger an aesthetic response in *one* of them. So obviously there is some fuzziness to this definition and exactly how inclusive it is. This is unavoidable. Like all definitions it has fuzzy edges.

exquisitely designed and crafted chair in our dining room or even the clever eye-catching packaging on an item in our pantry. I think the discomfort one might feel with my inclusive definition stems less from its general inclusivity than from the implicitly positive connotation of the term "art". When someone asks, sarcasm dripping, "You really would call *that* art?!" I think they're really asking if I think it is good art—a very different question. I have no problem accepting as art, for example, the latest three-chord single from some currently mega-popular but talentless rock group, because obviously a large number of people are responding to it as art. One can see evidence of this every day on any city bus: the rapturous expressions on teenagers listening to their IPods and other portable music players. It is very important to realize that because something *is* art does not mean it can't be *bad* art. The evaluation of a work of art is another issue entirely, one to be considered briefly later on.

Another, second, quite reasonable objection to my working definition of art is that it is circular: if art is what evokes an aesthetic response in at least some people, so what then is my definition of an aesthetic response? All I can reply to this is that an aesthetic reaction—i.e., being 'moved' or 'touched' by contemplation of a painting, a musical composition, a film, a poem or whatever—is very close to a universal human experience, so there is little practical need to define it. We all, or nearly all, know what it means to be affected by a painting, or a piece of music, or a poem. About this experience there is little dispute, unlike the question: "What is art?" So it makes sense to escape the ambiguity of the term "art" by moving to a definition that calls upon the experientially defined, and less ambiguous, "aesthetic experience." Again, I think any discomfiture really stems less from the definition begging the question than from the positive connotation of the word 'art'. Understandably a serious poet will choke at the thought of equating a sophisticated reader's empathetic response to her well-crafted work with some hormone-befuddled teenager's response to the saccharine lyrics of a top-ten love song. But the fact is both are aesthetic responses, and just as there is good art and bad art, there are profound aesthetic experiences and superficial ones. In fact, it may be that the differences in depth, or the nature, of the typical appreciators' response to a work of art is precisely what distinguishes the good from the mediocre from the bad.

The third major problem I see with my definition is the one that has particularly plagued the visual arts since the invention of photography—and now has reared its head in the area of artificial intelligence. It might reasonably be labelled "the found art conundrum". Perhaps its most dramatic representation and confrontation is Marcel Duchamp's infamous urinal. This artwork by the father of conceptual art did more to challenge thinking about art (or at least visual art) than any other 'creation' before or since. The story is familiar to any artist or student of the arts. In February of 1917, The Society of Independent Artists in New York City held an open exhibition. The fascinating and enigmatic Marcel Duchamp made a clear, albeit implicit, statement of his definition of art as "whatever the artist chooses to call art" when he installed as his contribution to the show a work named "Fountain"—a decidedly ugly urinal which he had 'signed' "R. Mutt"* and stood on end to add a phallic connotation.

I reject Duchamp's implied definition† as useless because it is so contrary to popular conception—in fact, deliberately antagonistic to the conventional concept of art. (This was intentional, of course, because baiting the bourgeoisie was the most popular sport among artists of the time). Artists are not magicians who can turn anything into art by calling it art. Aside from the alienating *hubris* of any artist who defines art as that which he chooses to call art (a *hubris* just as extreme as that of the philistine who defines art as only what he personally likes), this arrogant definition excludes from the set of things called art all the wonderful music and sculpture and narrative in cultures where there isn't even really a conception of 'artist' or 'art'.

However, I do accept Duchamp's urinal as a work of art. It does not matter to me that he, Marcel Duchamp *The Artiste*, chose to call his piss-pot art. I don't think the artist's intention is relevant to defining art. I may intend to create a work of art, but it is only accepted as a work of art if it produces an aesthetic response in someone: if everyone who looks at my creation just shrugs, my best intentions mean nothing. On the other hand, I may have no intention of creating a work of art when I take a casual snapshot that when shown to others elicits comments about how beautiful and moving the image is. Why I—and I'm sure many other people—sincerely do respond to Duchamp's urinal with what could justifiably be called an

---

* Apparently a character in a popular cartoon of the time was named Mutt.
† Duchamp was always reticent to make any explicit or direct statements about art, preferring his work to speak for itself.

aesthetic response has nothing to do with his intention or with his actually physically creating the object.

.

Duchamp didn't make the urinal: he merely found it, named it, flipped it over, put it on a pedestal, and signed it with a pseudonym. His gesture was a natural extension of what the photographer does: the man with a camera does not 'make" the striking image that he would call art; he 'merely' mechanically 'captures' on film what is already there.* Only God can make a tree, so trees cannot be art by the proffered definition. But then does an 'exact' two-dimensional copy of this tree on film have any claim to being art?† And Duchamp's urinal, while not created by God, but rather by the collaboration of some anonymous urinal designer‡ and factory worker, would not normally be considered art—because it is extremely difficult to imagine it eliciting an aesthetic response outside of the historical and cultural context in which Duchamp embedded it.

.

I think the way around this apparent conundrum is to admit that a thing is more than its physical "thingness": it is almost magically enriched by the experience of the person experiencing it. (Note the 'magic' here is in the mind of the beholder, not in the hand—or mouth—of the artist.) In the case of Duchamp's urinal, it is the whole historical and cultural context that those confronting it bring to it. Art is never free of external referents. If no one ever came to his urinal without being already equipped with the contextual knowledge that makes it significant, no one would respond to it— and it would just remain a urinal. Imagine a world where no one understood English and no Rosetta Stone existed to decipher the language. Would *Hamlet* be art? I admit my definition would indeed deny it the status of art, as it would also deny that honour to Beethoven's last string quartets were all human beings deaf. This might make some uncomfortable, but I think this exclusiveness is a small price to pay for the usefulness of the definition and its

---

* Of course I know the photographer frames it. and chooses certain camera settings, and then later modifies it in his darkroom or in PhotoShop. I'm not denigrating photographic art. The point is that the original image is out there, *found* in the real world.
† Of course, a photographic image is not really an "exact copy", a topic to be considered later.
‡ It is illuminating that another term for found-art is a "ready-made", as if the urinal popped into existence and the petit bourgeois designer of it and the proletariat factory worker who made it were of absolutely no importance—unlike the artist with his magic powers to give this humble object deep significance.

congruence with what both the sophisticated and the naive mean by the word *art*. It also balances to some extent the otherwise very inclusive nature of my definition.

CREATIVITY: MYSTERIOUS MEANS OF PRODUCTION

Having presented working definitions for science and art, the products of human *creativity*, it is time to be explicit about how that multi-faceted word will be used in the remainder of this book, and also look at how it is used by those others who have tried to understand it. Here are four common and fairly reasonable definitions of the word.*

- Creativity as unusual approaches to life and problems
- Creativity as a rich and complex association of cognitive elements
- Creativity as that mysterious thing artists (and perhaps scientists) possess
- Creativity as a process that leads to a culturally worthwhile creative product

The first definition is perhaps the most common one: creativity as unusual approaches to problems. This is clearly the definition being used by the authors of books with titles such as "Creative Child-Rearing" or "Creative Divorce". It is also, although perhaps to a lesser extent, the meaning for people such as DeBono (originator of "Lateral Thinking") and Osborn (originator of "Brainstorming") who both claim the ability to train people to "be creative". When DeBono holds one of his high-priced workshops for business executives, his intention clearly is not to turn these business-suited folk into artists or scientists. He simply wants to make them more effective problem-solvers—and thus more successful in what they do—usually making money. (Whether the creativity trainers' techniques are in fact useful to artists and scientists is another question—one that will be addressed eventually.) The concern of this book is not with this general thing—new, unusual and effective problem-solving techniques, although the relationship of problem-solving strategies to contributions to art and science will be examined, as will the claim that 'creativity' is itself a myth and only a word used to describe intelligent problem-solving.

---

* These four categories are from a 1988 essay by Calvin W. Taylor entitled "Various Approaches To And Definitions Of Creativity" which can be found in R.J. Sternberg's anthology *The Nature of Creativity* published by Cambridge University Press.

The second definition, that of creativity as a particular characteristic each person possesses to a greater or lesser degree, rather like a different kind of intelligence, is an interesting one and one that many psychologists seem to embrace. Intelligence has been labelled "convergent thinking" and creativity "divergent thinking"—and the relationship between these two in any person considered accomplished in his or her chosen field or in any exceptional problem-solver is a fascinating one—as is the interplay of heredity and environment in producing such a person. We have IQ tests that purport to measure "intelligence" (and which are moderately effective in predicting academic success); there are also tests to measure some equivalent Creative Quotient, a "CQ" , although the predictive powers of these tests fall far short of those of intelligence tests—a topic to be examined in the next chapter. So although the scope of this book definitely includes a look at what characteristics distinguish an Einstein or a Picasso from the mass of humanity, I will not use this definition as my operational one.

.

The previous two definitions can be simply expressed as "creativity as a learned skill" and "creativity as a type of intelligence". The third definition is more focused on the nature of the creator, on the gifts with which the artist or the scientist is blessed. Creativity, by this definition, is that property those who create art and those who create science possess. Those who are not artists and scientists (or at least professed artists and scientists) are excluded from the club. There is a bit of snobbishness in this definition, a snobbishness that I sheepishly have to admit to sharing, although I don't think this definition the best.

.

My interest, and the focus of this book, is on what lies behind humankind's invention of civilization—for as I've already said, I think it reasonable to define civilization as art and science.* So the way I will use the word creativity most closely matches the last of the four common definitions: Creativity is a process, but specifically and only that process which produces a worthwhile cultural product—a work of art or a better scientific understanding of our universe. (I will not call "creative" an unusual and effective way to handle separation from your spouse.) So while it may seem that this is the narrowest of

---

* I say this again without implying that such things as social justice or equality of opportunity are not important. Nor do I mean to suggest that quality of child rearing or that the treatment of the disadvantaged are not reasonable yardsticks by which to judge a civilization. It is only that these are not the yardsticks that I will be using. Usually there is very strong correlation between all these measures.

these four common definitions, nevertheless it will be the concern of this book and the one I will use—although I will have to repeatedly deal with these other definitions, since they so often are those used by researchers in this field.

.

I would also argue that my preferred definition is not really narrow at all, for creativity in this sense might include elements of all of the previous definitions. What I will mean by 'creativity" from this point on, unless otherwise indicated, is *that mysterious thing that is the impetus, the 'explanation', of humankind's unique creations: art and science.* Part of it may be some special characteristic of the human brain, a neurological gift bestowed only on some—or at least on some more than others. Part of it may be a personality characteristic (or even defect) in certain individuals that cause them to make things (e.g., symphonies or theories) that somehow matter deeply to the rest of us. Part of it may be a skill, a set of problem-solving strategies or a way of training one's eye or brain to attend to what is usually ignored—a skill that can be acquired to a greater or lesser degree by almost all human beings. I'll try to present objectively the evidence for all of the above, just as I will try to show how making art and making science are inter-related, inter-dependent, and how both are subject to influences, secret agents, which scientists, artists and their appreciators rarely acknowledge.

CASE STUDIES: ISAAC NEWTON AND JOHN KEATS

Consider these two. The archetypal scientist and archetypal poet: Isaac Newton and John Keats, two towering geniuses who stand at the gate to the modern world of science and art. Newton has been rightly credited with having ushered in the Age Of Reason and being the father of modern physical science. Keats, while he may not stand so alone as Newton, is certainly one of the major initiators—and arguably the most brilliant—of the Romantic Movement in the literary arts, a movement often considered a reaction to the Age Of Reason that the Romantics felt had dominated the intellectual landscape for almost a century. Newton lived to a ripe, cantankerous old age in the second decade of the eighteenth century. Keats died at the age of 25 in the second decade of the nineteenth Century. Newton was a nasty bit of business, by all accounts a totally disagreeable man. Keats, on the other hand, seems to have been a kind and decent young man with a sweet disposition. Both, however, had the experience of having their first major creative accomplishments ridiculed. They make a striking contrast and illuminate some of the contradictions inherent in doing art or doing science and thinking the other is one's enemy

.

Keats, born 1795, was the eldest son of a livery-stable manager. After his father died in 1804, his mother remarried, but the union failed. The family then moved from London to the nearby smaller community of Edmonton to live with Keats' grandmother, who subsequently succumbed to tuberculosis, the same disease that was to kill the poet eleven years later. It's likely that these familial losses affected the poet's themes.

.

Keats was educated at Clarke's School in Enfield, where he demonstrated early his literary talents by beginning the formidable task of writing a translation of Virgil's *Aeneid*. Although he read widely, it is interesting that his formal education was in the direction of science, medical science: In 1811 he was apprenticed to a surgeon-apothecary. He moved to London in 1814 to continue his surgical studies as a student at Guy's hospital. That same year he wrote his first poem, "Lines in Imitation of Spenser". In 1816 he became a Licentiate of the Society of Apothecaries and worked as a dresser and junior house surgeon.

.

But then the poetic urge overwhelmed him, and he devoted the remaining five years of his life to poetry, creating in that short span some of the finest English language poems of the nineteenth century, profoundly infused with a sense of the magic and mystery and beauty of life. His reputation as a poet seems to rest on his 'mature' poems, those written when he was ill (and so poor that he couldn't afford to marry Fanny Brawne, the love of his life), poems that seem to indicate an obsession with death and decay and despair. But his earlier poems, written while he was still in good health, do not fit the stereotype of a tragic, sensitive—even neurasthenic—soul. In the short time between devoting himself completely to poetry and becoming critically ill, Keats was a robust, vigorous young man passionately apprehending and appreciating the wonder of life—a person who more fits the stereotype of the young scientist than the tragic writer. And even later, when poverty and *consumption* (that incredibly apt word for tuberculosis) was sapping his strength, his poems still remain full of youthful vigour and enthusiasm for life— albeit coloured with a sense of the fundamentally transitory, autumnal nature of the most intense beauty.

.

His famous lines "When I have fears that I may cease to be / Before my pen has gleaned my teeming brain" speak less of despair than passion for life and creative expression. A poem from the Canadian Irving Layton has these equally memorable words: "Death is a word for beauty not in use." I think Keats would've agreed. I think most scientists would agree. I think Newton would've agreed.

.

So how could gentle John Keats lift a glass to curse Newton*, who was so clearly similarly engaged in trying to apprehend the mystery and beauty of the universe? How could Keats seriously accuse Newton of "unweaving the rainbow by reducing it to a prism"? Does not a poet unweave the rainbow too when he dissects it, takes the brilliantly coloured strands and weaves them into a tapestry of verse? In fact is not that unweaving of the rainbow what all artists do: take apart what nature has made so as to put it back together in an even more pleasing composition? Aristotle would say so.

.

"'Beauty is truth, truth beauty,'—that is all / Ye know on earth and all ye need to know." This Keats wrote in his "Ode on a Grecian Urn". Is this anti-intellectualism, a call for ignorance, for not

---

* Keats allegedly joined several other Romantic poets at a party in this boisterous condemnation of Newton and what they believed he stood for.

attempting to apprehend beauty? It has been interpreted that way, but a careful reading of the poem and a thoughtful consideration of the poet's life refute this interpretation. There is no question that Keats shared to some extent the naïve fears of his romantic contemporaries that science might destroy the mystery of the world. However, Keats was a young man whose understanding of the new science of physics may have been limited, but whose understanding of natural science we can assume was substantial, given his training in medicine and the evidence of careful observation of the natural world in his poetry. So I think it reasonable to interpret the famous equation of truth with beauty as an insight into the role that beauty plays, not just in the arts, but also in the search for truth elsewhere—including in science.

Newton, we can reasonably assume, knew that there was beauty in truth. The beauty and elegance of his theoretical contributions still awe physicists. If they can see this aesthetic perfection, surely the creator of these theories was also aware of it.

Isaac Newton was born in 1642 at Woolsthorpe, England, where he attended school. In 1661 he began attending Cambridge. Five years later he was elected a Fellow of Trinity College, and in 1669 became Lucasian Professor of Mathematics. He remained at the university until 1696, and during his years there was (by his own evaluation as well as that of others) at the height of his creative powers. And a dizzying height it was: he is considered by many, perhaps most, scientists to be the greatest scientific intellect ever. He stands in relationship to science as Shakespeare stands in relationship to English literature. He not only discovered the fundamental principles of physics, but he also to a great extent defined its methodology. And if this weren't sufficient accomplishment for a mortal, he invented (or should it be 'discovered'?) calculus, that most fundamental mathematical tool, to explain and defend his physical theories!*

What is particularly relevant to understanding the paradoxical nature of creativity is something that is considered by many to be skeletons in Newton's intellectual closet, things best tactfully left unmentioned. Still the fact is that Newton spent a large portion of his life experimenting with alchemy and was also profoundly religious. He disliked his own theory of gravity because it was too "mystical": the

---

* Leibniz did so as well, about the same time, and their acrimonious conflict over priority is infamous.

idea of action at a distance bothered him. (As it did Einstein—who resolved it with his General Theory of Relativity.) Yet Newton himself was more than a bit the mystic. Deeply religious, he venerated the Bible as the Word of God. Throughout his whole life he studied and wrote about theology. (He wrote a book on Judaeo-Christian prophecy, the deciphering of which he thought essential to the proper understanding of God.)

.

These facts embarrass those who see Isaac Newton as the ultimate incarnation of The Scientist—pure rationality personified. In contrast to this, the biographical facts suggest a rather 'romantic' figure—an alchemist and a mystic. While there is no evidence of a particularly strong interest in the conventional arts, there is plenty of evidence for a life-long engagement with the same concerns that are associated with the romantic artist—spirituality and the underlying unified beauty of the universe. As the perceived enemy of passion and mystery, Newton was framed, was set up as a straw man to attack by some.

.

It *is* true that Newton did quite literally unweave the rainbow—but only after he had first created it! The famous experiment central to his *Theory Of Optiks* involves taking a beam of white light and passing it through a prism, thus breaking it down into its components—the familiar spectrum of the rainbow. Newton and his prism unwove, not really the rainbow, but rather white light. Unwove it *into* a rainbow, and then took the multi-coloured stands and passed them through another prism to weave them back together again into white light. This demonstrated the principle that white light is composed of all the colours of the rainbow—a very radical idea at the time.* It is sad that such analytic and synthetic magic was perceived by the Romantics as endangering the mystery and wonder of the world.

.

Alexander Pope's couplet is apropos: "Nature and Nature's laws lay hid in night; God said, Let Newton be! and all was light." It is difficult to believe that Keats would've preferred darkness, no matter how mysterious.

---

* His contemporaries believed that colours were modified forms of homogeneous white light.

OF TWO MINDS ABOUT SPLIT BRAINS: IQs AND CQs

"I'm of two minds about creativity and intelligence: one half believes they are related, the other half doesn't—and I'm not sure which half is more intelligent."

—Hippokrites

"Talent without genius isn't much, but genius without talent is nothing whatsoever."

—Paul Valery (*Cahiers*)

A tonic clonic (or *grand mal*) epileptic seizure is a disturbing thing to witness. You're talking to this perfectly healthy-looking man when suddenly he releases an involuntary scream—the result of a sudden, severe contraction of the muscles that control breathing.

This is followed immediately by his eyes rolling up into his head as consciousness is lost. His face flushes red, his breathing stops, he collapses to the floor and his back arches. Then his entire body begins jerking about wildly as his (previously) voluntary muscles fire randomly; even antagonistic muscles such as the biceps and triceps may attempt to contract simultaneously, resulting in a violent quivering as they pull with full force in opposite directions.

All this is in response to illogical signals sent from the motor cortex of the brain to the neuromuscular junctions.

The seizure may last for several minutes and sometimes, in extreme cases, start again shortly after it ends. The immediate physical cause of this horrible and horrifying event is the random, erratic firing of cortical neurons, triggered by stress, drugs, repetitive sounds, flashing lights, or even touch to certain parts of the body.

Epilepsy has often been associated with madness and possession—and creativity.* That these are so frequently linked is a theme that will be explored in detail later.

---

* Feodor Dostoevsky is perhaps the most famous epileptic.

.

What is relevant here, however, is the strange chain of events that began with a radical treatment for this disease—and ended with the current pop-science foolishness about so-called 'right brain thinking'.

RIGHT BRAIN, WRONG BRAIN

Currently epilepsy is usually treated—and well controlled—with diet and anticonvulsant drugs such as diazepam. But in extreme cases surgery has been resorted to, including the drastic measure of severing the corpus callosum, the structure that connects the two hemispheres of the brain. The rationale for this extreme measure is that it aborts the random bursts of neuronal activity that amplify as they travel outward from their origin, wreaking spectacular havoc as they wash over the motor cortex—that part of the brain that sends signals to the muscles. As with a breakwater, the rising waves of random neural activity are stopped cold when they arrive at the neurological channel that connects the two hemispheres—this now-severed corpus callosum.

The side effect of this surgery is a research opportunity. People who have had their corpus callosum connection cut become people with two almost independent brains. Professor Michael Gazzaniga, from Dartmouth College worked with "split brain" patients for over 30 years.* In a BBC documentary† he summarized some of the findings—findings, misinterpretation of which has led to the simplistic idea that the left hemisphere is responsible for intelligence and the right hemisphere for creativity.

A typical "split brain" study involves presenting different visual stimuli to the different hemispheres of a person who has had his commissural pathways severed. This can be arranged by separating with a partition what is seen by the subject's left and right eyes.‡ From such studies a number of interesting things about hemispheric specialization have been discovered, and the century-old theory that verbal skills tended to be handled primarily by Broca's area and Wernicke's area, both located in the left hemisphere, and that spatial reasoning was primarily a right brain specialization were confirmed. As Professor Gazzaniga describes it: "Take a split brain patient and measure their preoperative IQ and problem solving. Then you disconnect their hemispheres, and go back and measure the left

---

* R. Sperry is another major figure in split-brain research; he received the Nobel Prize for his pioneering work in this area.
† Broadcast on BBC's "Brain Story" series in 2000.
‡ This is somewhat more complicated than this summary suggests, because the information from the two eyes is 'mingled' when the optic tracts meet at the optic chiasma on their way to the visual cortex—so both hemispheres receive information from both eyes.

hemisphere's IQ and problem solving, and it hasn't changed a whit. The right hemisphere is, on the other hand, kinda {sic} dumb. The visual test produced the opposite result: The left hemisphere just fell apart—just couldn't do the test."

.

In the decades following the first split-brain studies in the 1960s, many people were tempted into extrapolating the established, but very specific, hemispheric differences into a grander—and fuzzier— differentiation and specialization of the two brains we all have tucked up in our craniums. Studies of stroke victims with localized cortical damage (another example of personal misfortune yielding up research opportunity) and studies of differences in verbal versus visual skills as well as personality differences between right-handed folk—presumably left hemispheric dominant—and lefties were recruited to support the simplistic idea that the right brain housed our creativity and spontaneity, while the left brain took care of the business of logical thought. Perhaps partially because of the heady, youth-dominated atmosphere of the sixties and seventies, the allegedly wild and crazy (and fun-loving) right hemisphere was favoured over the tight-assed, let's-be-reasonable left hemisphere.

.

So it shouldn't be surprising that suddenly books by educationists looking for a new twist and hack writers looking for a hook latched onto this idea, this equation of creativity with the 'Right Brain'. The books still are pouring out: *Free Flight— Celebrating Your Right Brain*; *Drawing on the Right Side of the Brain*; *Cooking With the Right Side of the Brain*; *Unicorns Are Real— A Right-Brained Approach to Learning*; *Right-Brained Children in a Left-Brained World— Unlocking the Potential of Your ADD Child*; *The Natural Rider— A Right-Brain Approach to Riding*; and one of my favourites, *Money 101— For the Creatively Inclined— Right-Brained Finance for Left-Brained People.**

.

Meanwhile, scientists, those poor folk mired in stodgy left-brain thinking, were pooping on this party. (Not that anyone in the educational system noticed.) Further research on split-brain patients and the development of brain-scanning technology (PET, CAT, and fMRI) produced overwhelming evidence that the two hemispheres, while indeed having some specialized functions, some asymmetry, are primarily backup systems for each other. It has become increasingly clear that what is most remarkable about the two hemispheres is their

---

* One has to wonder if a CA who practices "creative accounting" wrote this. If so, he's flagging himself for a visit from the Internal Revenue auditors.

similarity and plasticity! The rough and ready generalization that the left hemisphere is *usually* somewhat more specialized for verbal functions and the right for spatial functions is reasonable and well documented, but the association of right brain function with creativity is total nonsense.*

* Here is how one noted researcher in the field more diplomatically put this. "Currently [a relatively] popular belief is that creativity is associated with right-brain processes. There is, however, little relevant evidence to support this oversimplified theory, except possibly in the visual arts." (P. E. Vernon, 1987, in D.N. Jackson & J.P. Rushton's *Scientific Excellence* published by Sage.)

IF NOT TWO BRAINS, MAYBE TWO MINDS: ONE INTELLIGENT AND ONE CREATIVE

One issue split brain research and its misinterpretation has highlighted is the assumed distinction between intelligence and creativity. This distinction and the question of its validity is one that has to be addressed early in any discussion of the nature of creativity. So, given that the right brain doesn't 'do' creativity and the left 'do' intelligence, is there any reason to assume these are really two different tasks at all?

It is tempting to say that the idea of intelligence predates the idea of creativity, but this may not be entirely true. For example, in ancient Greece the creative artist—poet or musician—was considered to be inspired, delivered of his work, by a Muse. I don't know of any evidence that a genius such as Sophocles was considered particularly intelligent, at least in the way that someone like Archimedes or Plato was considered intelligent. As is well known, Plato didn't much trust poets as being intelligent enough to even be tolerated in a Utopia, never mind run the place.* A Philosopher King was considered a good idea because a philosopher's accomplishments were based on skill at reasoning. Poets just had their goods delivered by the Muses.

It is also worth noting that most Greeks tended to equate intelligence with knowledge, with a prodigious memory of acquired information, even more than with reasoning powers, skill at argumentation, or problem solving skills. This makes some sense. There was no Google search engine available. Even book publication was more than a millennium in the future. So the person who simply *knew* and could remember a lot of stuff was considered intelligent.

Even when we look at the first scientific efforts in the Nineteenth Century to understand eminence or genius in the field of art and science, this distinction between creativity and intelligence didn't seem to be considered. In this book I have chosen to look at eminence—substantial creative contribution to what we call civilization—as an unequivocal example of creativity, because it clears the discussion table of confusing clutter.

---

* I know this a bit unfair and will annoy Plato scholars. Plato seemed most concerned about the power of poets to manipulate the emotions—at the expense of reason.

In the first scientific, or quasi-scientific, attempts to understand eminent achievement, the distinction between creativity and intelligence didn't play a role, and before 'intelligence testing' the convention was to simply call accomplished individuals 'intelligent'—no matter the domain of their contribution. So, were Newton or Keats 'creative' or 'intelligent' or both?

IN THE RIGHT WING / LEFT HEMISPHERE WE HAVE IQ WEIGHING IN AT 150 POINTS

One of the common definitions of creativity (one I've chosen not to embrace, but one which I will have to deal with occasionally) is that of a talent or gift somewhat like the common current conception of intelligence—perhaps even a subset of that general thing called intelligence.* Many of the psychological studies of creativity take some variation of this as their operational definition, and a sometimes fierce debate rages over whether this thing called creativity is separate from intelligence, maybe even in some sense its antithesis, or whether it is just a type of intelligence. This debate is confounded by confusion about what precisely is meant by intelligence.

.

Francis Galton, Charles Darwin's cousin and a prominent nineteenth century scientist, is the individual most responsible for our current conception of intelligence. He believed that genius was defined by exceptional intelligence, an inherited trait. He also made the first attempt to find a way to measure, quantify, this trait called intelligence. His book, *Hereditary Genius*, is seminal to the whole field. Along the way he coined the phrase "nature versus nurture" for the ideological conflict that remains to this day the most hotly debated issue in the social sciences. He also refined the important statistical tools know as linear regression and correlation which quantitatively measure relationships between two variables. On the basis of his beliefs, and correlations he found, Galton believed nature, our genetic inheritance, to be of primary importance in determining our intelligence. So he founded a popular eugenics movement to improve the human race through selective breeding. Although Galton was certainly well intentioned, the subsequent abuses of eugenicists— extending from sterilization laws being passed in the 30's in the United States to Nazi "ethnic cleansing"—have, most unfortunately, over-shadowed in the public mind his many substantial contributions to civilization.

.

Galton, undoubtedly influenced by his cousin's *Origin Of The Species*, initially believed that intelligence was simply an inherited trait— although in later years he was more willing to admit the effects of

---

* For example, R.W. Weisberg argues, in his book *Creativity: Beyond The Myth Of Genius*, that creativity is nothing more than logical sequential problem solving.

environmental factors. Intelligence thus might be considered analogous to height. Our adult height is, excluding the usually minor environmental effects of diet, a direct result of the genetic hand we are dealt by our parents playing in their conjugal bed. There is even a formula that can give fairly good estimates of the probable heights of our offspring. It involves averaging Dad's height and Mom's height, while giving more "weight" to the parent having the same gender as the offspring—plus a "regression to the mean" factor. This last is an expression of the tendency for the genetic average to "pull" things back to what is typical for the population. If, for example, my mother is 5'10" and my father is 6'2", one might expect me to be a bit more than 6 feet tall, but since both my parents are far taller than average, I might very well be not even as tall as my mother.

Galton believed, as have many others since, that inherited intelligence works very much the same way. Brilliant parents will most likely produce a child who is very bright, but nevertheless probably with a brightness shining somewhat less intensely than the average of their own brilliance—because of this statistical 'gravity' tugging the next generation back toward the mean. Of course occasionally, again due to the nature of probabilistic events, the offspring might actually be taller (or brighter) than either of his (or her) tall (or bright) parents. If one accepts IQ as a measure of intelligence, there is some evidence that this is true. Unfortunately, the evidence is confounded by the fact that IQ as a measure of intelligence doesn't have the same commonsensical validity as does a yardstick as a measure of height.

What is intelligence? It is what IQ measures. What is IQ? It is a measurement of intelligence. Do we have a serious circularity here, or what? Is not a question being begged?

Particularly relevant to this are two important concepts in scientific thinking called "reliability" and "validity". Reliability refers to the repeatability; i.e., the consistency of repeated measurements or observations. Validity refers to whether a measurement really measures what it purports to measure. For any scientific theory to be accepted, the observations on which it is based must meet both of these criteria.

I have encountered many people who say they do not "believe in IQ" because they don't think it is "reliable." They are confusing these two scientific terms—albeit understandably, because in common usage they are often treated as synonymous. But if I'm feeling in a difficult and pedantic mood, I reply that they are quite wrong: IQ is surprisingly *reliable*.* If I give a person a standard IQ test (e.g., the WAIS) at age 21 and retest him at age 61, these two scores probably will be very similar, differing by only a few points—unless sometime in those four decades my test subject had fallen off his bike and landed on his head or had some other trauma to his brain. The real debate about IQ and intelligence is not the reliability of the measurement called IQ: it is about the *validity* of IQ as a measurement of intelligence, as that word is normally used.

.

I have developed a new test of intelligence, a new adult IQ test: I call it the BIG (Best Intelligence Gauge). It is even more reliable than the WAIS or the Stanford-Binet!† I measure the distance from the bottom of a person's foot to the highest point on his or her head: The larger the number, the brighter the subject. Repeated measures give highly *reliable* results. (I scored a very respectable 6'1" at age 21 and my latest measurement comes up the same.‡) Of course, again, some trauma, such as decapitation, could cause a lowering of my BIG IQ. (I estimate that losing my head—something my wife imprecisely accuses me of almost daily—would reduce my IQ to about 5'1".) Of course, there are always critics out there in the sceptical world of science. They don't dare challenge the empirical evidence supporting the *reliability* of my test, so instead they question its *validity*. (Incidentally, many of these critics are women, probably because the average BIG IQ of women is considerably and significantly lower than that of men.)§

.

I think we are back to the issue of operational definitions. There is nothing inherently 'wrong' with my operational definition of intelligence as height: it just doesn't seem very useful or very congruent with what most people mean when they use the word intelligence—so it wouldn't be very tall of me to expect my definition to attain universal acceptance.

.

But, of course, what *is* usually meant by intelligence is not all that clear either—especially since it became more associated with psychometrics than accomplishment. Certainly 'intelligent' is not

---

* In statistical terms, age-repeated measures of IQ tend to correlate around +0.7, which is considered a "moderate to strong" correlation. (For comparison, this is roughly the same as the correlation one finds between height and weight.)
† These are the two standard IQ tests most commonly used.
‡ I may have lost a tiny fraction of an inch from gravity compressing my spinal cord.
§ I'm sure I'd get hate mail from short people.

quite as bad as 'creative' in terms of the plethora of meanings, but one important similarity these two words do have—and a problem with both that occurs in discourse about them—is that sometimes they are used to describe a person who has demonstrated the respective attribute by significant achievement—and sometimes to describe a person who merely shows certain superficial characteristics associated with the attribute. Art students, for example, are often automatically considered to be 'creative' and science students to be 'intelligent'. (And, thanks to the very mixed blessing of IQ testing, people with high test-scores often are automatically considered intelligent.) As with creativity, I think it sensible only to consider a person 'intelligent' if they do intelligent things. I can't understand how people who consistently behave stupidly, and never accomplish anything intelligent, can be reasonably called intelligent. "He's really smart; he just doesn't apply himself!" seems a damn silly thing to say. The proof is in the quality of the pudding—not in the attractiveness of the recipe.

The history of IQ testing and the concept of predictive validity are pertinent. To a great extent Galton understood intelligence in the same way I do: intelligence is an attribute of people who do intelligent things—make meaningful contributions to the world, to apprehending the world, to civilization, to art and science. IQ testing, on the other hand, was originally developed as a tool to screen students in the French education system. It was designed to test for the neurological (or perhaps learned) characteristics associated with a student succeeding at some particular level of academic work. In short, IQ was intended to predict academic success or failure.[*]

Enter, stage right, the concept of predictive validity. The problem with the whole idea of validity is that while it is easy to define it as measuring what you think you're measuring, the question of *what you think you are measuring* remains a thorny one. If what you think you're measuring is intelligence, you still have to say *exactly* what you mean by intelligence—other than it is what your measurement measures! If you're willing to say that intelligence is accomplishment of some particular sort (e.g., good grades in 'academic' subjects in the French school system), then you avoid begging the question, and you stand on solid ground. The reason this is so is that the validity of your

---

[*] Well, actually it was originally intended to predict academic failure, not success—to weed out those students who probably wouldn't be able to cut the mustard in an academic program. And, in fact, it works better at doing that than the other way around.

measurement can be tested empirically. If most, or a significant proportion, of those who score high on your IQ test do better in school than those who score low, you certainly have some justification for saying your test measures intelligence. This is called predictive validity.

So a fair judgement of the validity of intelligence tests must be based on its powers of prediction: does a high IQ score predict intellectual accomplishment? And what kinds of accomplishment? Merely good grades in school? An important contribution to science or art? A number of studies have tried to answer these questions, the most interesting and meaningful being longitudinal studies; i.e., studies of individuals over long periods of time. Do these studies support the now commonly accepted idea that IQ predicts accomplishment in the world outside of school?

The short answer is no, not really—or at least not exceptional accomplishment. The most comprehensive longitudinal study, hyperbolically called *Genetic Studies of Genius* was initiated by Lewis Terman in the 1920's.* While at Stanford, Terman designed the well-known and commonly used "Stanford-Binet" IQ test, which he based on Binet's and Simon's work. He took his test very seriously, operationally defining "gifted" or "genius" by exceptionally high scores on his test: His famous 50 year study followed the lives of approximately 1,500 children who by age twelve had scored 140 or greater on his new IQ test. A massive amount of data was collected on these individuals who Terman affectionately referred to as his Termites. He didn't just look at professional accomplishment, but also at a great variety of other variables, everything from marital status to general health. He dispelled the myth of high IQ kids being sickly and socially inept, for his Termites actually were healthier, more well-adjusted and successful in life than average. However, he failed to support the notion that his test could detect genius. It seems only a tiny handful made any notable creative contributions, and these were things like the invention of the army K-ration and the idea for the "I Love Lucy" sitcom. On the other hand, two kids who took the test but *didn't* make the cut, went on to win Nobel Prizes for their

---

* The full text of this five-volume work has been scanned and is available on the Internet. Another study which looks specifically at the high end of Terman's sample only further supports the notion that even extremely high IQ scores are predictors of creative accomplishment: David Henry Feldman's "A follow-up of subjects scoring above 180 IQ in Terman's genetic studies of genius" published in *Council for Exceptional Children* (1984 Vol. 50, No. 6, pp. 518-523.)

creative accomplishments in physics: William Shockley in 1956 and Luis Alvarez in 1968.

Ironically, Shockley, like Galton and like Terman (who flunked him!), thought eugenics was a good idea. His quite vocal support of selective breeding of humans got him, as it had Terman, in a lot of hot water and sullied their reputations. Interestingly Shockley's position on this can be taken as a refutation of Terman's belief in a 'g' factor; i.e., an overriding general intelligence. Once again we see ethical intelligence and creative intelligence are not necessarily correlated.

There are two less rigorously scientific exercises that anyone can conduct that support the findings of the Terman study.

Find out what those straight-A students you knew in high school are up to now. Or if your high school, like mine in pre-politically-correct days, had special classes called "Honours Classes" for those who managed to score high on a standardized IQ test, go and hunt down those class brains to see what has become of them. This exercise is surprisingly easy with the Internet. And it often offers the somewhat nasty pleasure associated with finding out that 'the most likely to succeed' is working in a factory for minimum wage, that your school's 'super-brain' dropped out of grad school and is living at home with his mother.

The other exercise involves finding a copy of the MENSA directory. MENSA is an organization whose members have only one thing in common: they all scored in the top 2 percent on a supervised IQ test. Members are listed in a massive directory so that they can contact each other—and, presumably, say really intelligent things to each other. Then start trying to find within this high-IQ phonebook any notable (living, of course) artists, writers and scientists whom you admire. Years ago, I tried this exercise for an hour of so. I did find Isaac Asimov listed, but not a single other artist or scientist I searched for.

Now I'm fully aware that the results of either of these exercises can't be taken seriously as a scientific refutation of the predictive validity of aptitude testing. (Who really would expect many intelligent and accomplished grownups to waste their time taking a MENSA entrance exam?) Those wishing to look more closely into the ability of aptitude tests to predict real world accomplishment can search out the scientific literature. But what they will find is that IQ is somewhat—but far from strongly—correlated with purely academic accomplishment. However, then they should also look at how well academic accomplishment correlates with actual intellectual or artistic accomplishment—or if they have nerves of steel and a high tolerance for prideful idiocy, attend an academic senate meeting at any university.

.

The bottom line is that putting a lot of faith in IQ as really having much to do with actual achievement is rather—well, unintelligent.

IN THE LEFT WING / RIGHT HEMISPHERE WE HAVE CREATIVITY AS A FEATHER-WEIGHT

What about creativity? I think it is fair to say that those with more limited definitions for intelligence, such as "that which predicts academic success", are those most inclined to feel a need for another term to describe innovative accomplishment. For them, intelligence needs be complemented with something else—and 'creativity' seems a good word—to account for the accomplishments of an Einstein or a Picasso.

Consider the radical shift of perspective and accompanying insights of Einstein's Special and General Theories of Relativity. Who would deny that Einstein was intelligent *and* creative? And who would deny that Shakespeare was creative *and* intelligent? But if these, in fact, are two separate things, what is their relationship to each other? Can you have one without the other, like love and marriage, or are they most useful fitted together, like horse and carriage.

One common take on this is that a certain minimum of intelligence (in the sense of being fluent at critical, logical thinking) is prerequisite to creativity. If you don't have enough neurons to rub together, you're never going to get a spark. Still, so the argument goes, once this minimum standard is met, there is little relationship of creativity to IQ. An IQ of 180 does not a Michelangelo or Einstein make, but no moron is ever going to paint a Sistine Chapel or redefine the concept of gravity.* Furthermore, the entry level IQ varies from one field of endeavour to another. Visual artists can get by with less IQ type intelligence than novelists who can, in turn, get by with less than physicists or mathematicians.

There is undoubtedly some truth in this view of things. Some ability at critical thinking, crudely estimated by IQ, is almost certainly prerequisite to creative accomplishment. The writer who cannot edit isn't a writer: he's a scribbler. (Apropos, Truman Capote, when asked what he thought of the writing of Jack Kerouac, replied that it wasn't writing: it was typing.) The importance of critical thinking is certainly greater in science with its demand for empirical validation than in the arts, but it is still of huge importance in the arts. The writer has to

---

* Well, almost everything has an exception. Gore Vidal once remarked of Andy Warhol that he was the only genius to boast an IQ of 60.

revise and organize; the painter has to compose—and *dispose* of what does not fit.

.

In my youth, a youth that did not include any personal acquaintance with artists or scientists, I naively expected accomplished writers and artists and scientists to be brilliant lights. So it came as a rude awakening when I actually started meeting creative and intellectually accomplished people. Many of them were blinking idiots—outside of their chosen field of endeavour. Dylan Thomas reminisces in a brilliant monologue introducing one of his poetry readings about this disillusionment, remarking what a shock it was for him to observe the poet Mathew Arnold, author of some of the most dignified and resonant poems of our time, at an adjoining restaurant table, and the great man "spoke and laughed much too loud." The more one gets to know creative people, be it through actual contact or through biographies, the more one wonders how the hell they managed to create what they did. Visual artists, in particular, often seem to be total flakes. When I've served on selection juries for art exhibitions, I am constantly re-amazed at the banal, even downright stupid, 'artist statements' that accompany the submission of first-rate art.

CONVERGENT AND DIVERGENT THINKING: RIGHT ANSWERS AND NEW ANSWERS

One important factor that has subtly, inadvertently, contributed to the belief that creativity and intelligence are two separate attributes is the nature of the tests that have been developed to measure them. When most people think of a 'test' they think of a series of questions, each having just *one right answer*.* No one makes it through formal schooling without taking literally thousands of tests of this type. And if a question on such a test does happen to have more than one right answer, we consider it flawed. Tests of this type are called 'congruent'. IQ and other aptitude tests are said to be tests of congruent thinking. Your IQ score depends on how many times you successfully pick from a set of possible answers the one that the test-maker considers to be correct.

I had an interesting email discussion on the nature of these types of tests with my daughter when she was preparing to take the General Abilities portion of the Graduate Record Exam (a test required of all applicants to graduate school in North America). Her field is pure math, and in that section of the test usually there is no ambiguity about which alternative answer is correct, but the verbal questions disconcerted her. My advice to her was to use 'common sense' to guess which alternative was the one most obvious—and to suppress any tendency to look for subtle relationships. For example, I said, consider a question that asks "which of these does not belong" and then presents the following list: apples; pears; bananas, turnips; grapes. As a poet, I can't help but notice that only 'apples' begins with a vowel, but common sense tells me the approved correct answer is 'turnips' because the rest of the items are fruits.

She replied with the following. "It's maybe going far afield here to choose the one that has a vowel in the beginning, but perfectly reasonable would be 'bananas' because the rest are round or because it has an inedible peel; or else 'grapes' because it's on a different scale than the others. The only reason you choose fruit is because you imagine the test takers are looking for categorisation rather than characteristic." Her answers are quite defensible, as is, I feel, my

---

* Actually there is no reason other than convenience in scoring that such tests have only one 'correct' answer. What really matters is that some answers are indisputably *incorrect*.

choice of 'apples', but one is never given a chance to 'defend' one's answers on such a test. Her defence of these alternative answers shows some skill at critical thinking and recognizing relationships, but were she to have chosen 'bananas' because she found *two* perfectly reasonable differences between it and the others, she would be scored the same as an illiterate who chose 'bananas' at random. Those who see a variety of possible 'right' answers based on different relationships and criteria have to demonstrate their 'intelligence' by knowing which is likely to be the one that the test-designer, and most people, consider 'right'; i.e., what is likely the most common 'right' answer.

.

Because it has come to be assumed that a major component of creativity as an attribute is the ability to see *uncommonsensical* relationships (relationships that most people don't see), tests without right answers were developed, tests that didn't measure convergence onto a right answer, but rather divergence from the usual or commonplace answer. They were called tests of divergent thinking.

.

For example, one such test involves coming up with unusual uses for a common object, such as a paperclip. The more responses and the more unusual the responses, the more points one earns on this test. Another such test involves what are called 'remote associations'. Once I was visiting an elementary school as part of a "Poets in the Schools" program, and they put me in a Grade Three class. In an attempt to get them to understand the idea of a simile, I had them try finding what psychologists call "remote associations" with a little game where one tries to find unusual similarities between diverse things. Everybody thinks of an object. Two people are randomly chosen to say aloud the object they were thinking about, and then everybody tries to find ways in which these two things are alike. One little girl said "camel" while a boy suggested "motorcycle". In a flash, another boy was waving his hand wildly about. "So how is a camel like a motorcycle?" I asked. "They both," he said grinning, "have gas."

.

Now I can't imagine where he learned that fact—if it is indeed a fact—about camels' tendency to flatulence, but I think anyone would have to agree his response was more 'creative' than saying they both are modes of transportation. The latter would have scored points for him on a convergent test, but his actual reply earns him high marks on a divergent test.

.

Various tests of divergent thinking will be examined in more detail in a later chapter on the uses and abuses of psychological approaches to creativity. What is of concern here is the question of the predictive validity of such tests. And, as is easily understandable, there are incredible methodological problems involved.

The first is the difficulty of objectively scoring any test that doesn't have 'right' answers: the predictor variable. The second is the difficulty of objectively evaluating the creativity of a person's life: the outcome variable. It is difficult to find longitudinal studies to support the hypothesis that results on divergent-thinking tests correlate with subsequent creative accomplishment. In short, the tests have little predictive validity.

One thing of primary importance, acknowledged in passing but largely ignored in trying to understand and resolve the conflicted relationship of convergent and divergent thinking, of intelligence and creativity, is that of personality, including motivation. One of the reasons that IQ or aptitude tests so often fail to predict eminence or even significant accomplishment, and probably the primary reason creativity or divergent-thinking tests fail even more miserably, is that they do not measure all the other secret agents—the individual's worldview, level of motivation, depth of passion, and any of those innumerable other features of an individual besides his or her ability to solve logic problems or invent unusual uses for a paperclip.

A second very important factor is that the distinction between convergent and divergent thinking is far blurrier than it appears to be. The equation of IQ with 'intelligence' in most of the ways the latter term is used is partially responsible for this. As Anastasi, a major figure in the field of psychological testing, points out, "For the general public, the IQ is not identified with a particular type of score on a particular test, but is often a shorthand designation for intelligence."* And she goes on to say that this is not merely a deplorable popular misconception, it engenders misunderstanding even for many professional psychometricians and is indicative of "a need to re-examine the general connotations of the construct 'intelligence,' as symbolized by the IQ." IQ tests estimate the level of very specific cognitive skills, and it is incorrect to assume an exact equation of these skills with 'convergent thinking'. There are other forms of convergent thinking. And divergent thinking includes more than unusualness of associations.

All this can be made clearer by concrete example. Consider two indisputably creative, intelligent individuals: one a scientist and the other a visual artist. Both of these will receive more detailed treatment in case studies later.

A Scientist. Reading a biography of Einstein is a corrective to any misguided confidence in psychological testing, academic performance, or any conventional indicators of either intelligence or creative potential. It also thoroughly confounds any simplistic distinction between intelligence and creativity.

---

* Anastasi's *Psychological Testing 6th Edition*, page 362.

Young Albert might even have been flagged as mentally subnormal were he to have been evaluated by a psychologist using currently accepted criteria for measuring the intelligence of children. Allegedly he didn't speak until he was four years old—something that on either the Gesell or McCarthy scales* for estimating intelligence would put him in the 'well-let's not expect too much of this Munchkin category'. His academic record was hardly stellar. He was notorious for cutting classes, and when he finally graduated in 1900 from the Swiss National Polytechnic, his professors wouldn't recommend him for a position in the university. (One of his professors even subsequently remarked on how mediocre were Einstein's mathematical skills—allegedly calling him "lazy" and a "dunce".) So Einstein didn't upon graduation slip into a nice tenure-track academic position at some 5-star university because of the brilliance of his academic career. No, he got his degree and landed in a dead-end job as "technical expert, third class" at a patent office.

While he clearly did demonstrate an ability to find relationships between apparently disparate things (such as being in an elevator in free fall and the experience of being outside of gravity in outer space—the inspiration for his General Theory of Relativity), he didn't, in any of the many post-fame recorded conversations, show any particular penchant for making unusual associations. One suspects that if he were asked how a camel and a motorcycle were alike, he would say, like most people, something about them being useful for transportation.

And although I know of no record of him ever taking a conventional IQ test†, one can't help but suspect that, despite his obviously unquestionable genius, he probably wouldn't have performed that much above average. (A physicist that many consider almost Einstein's equal in intellectual stature, Richard Feynman, *was* given an IQ test as a young man and scored a mere 124 points, only better than about two-thirds of the general population!) In fact, young Einstein when he first took the entrance examination for the Swiss Polytechnic (which really was just a glorified teachers' college), got results so poor that he felt he had to take a year off to attend the "Kantonsschule" in the small town of Aarau in order to upgrade!

---

* Two common scales for estimating intelligence in young children.

† An IQ score of 160 (high but not *that* exceptional) keeps surfacing in Internet discussions of this, but I've found no concrete evidence that he was ever actually tested.

In the later years of Albert Einstein's life he was portrayed as a man of phenomenal, superhuman intellectual and creative abilities, capable of thinking thoughts far beyond the comprehension of 99% of the rest of humanity. His silliest remarks about politics were taken as the deepest wisdom.* And after he died, a few scientists seriously thought an analysis of his pickled brain could explain his genius.

Albert Einstein certainly deserves to be called a genius, one of our most creative and intelligent men. He deserves this because of his achievements. What is instructive about a closer look at this man is that without his real intellectual, creative accomplishments, there really is nothing particularly remarkable about him. His high IQ (and presumably high CQ) are only attributed to him *ex post facto*.

An Artist. Next let's imagine the great surrealist, Rene Magritte, as he walks into his studio and confronts his easel, on which sits a blank, primed canvas, and thinks about what he is about to do and how he approaches the task. He wants, or needs, to make something—a painting that will elicit an aesthetic response in him and, he presumably hopes, in others. The construction of this painting will require a variety of skills, some learned, some inherited—and some classifiable as convergent, others as divergent.

Being a surrealist, he will probably first have to imagine something representational, a scene, a bizarre scene imbued with some inexplicable emotional resonance. This is clearly divergent thinking, for there is no correct answer as to what scene he should imagine, but is there any test of divergent thinking that can test this skill? The answer, at least as far as I know, is that there is not. Nor would it seem likely that Magritte would necessarily shine on a test that measured how many unusual things he could think to do with a paperclip.

Then he would set to painting the scene, surely adjusting elements of the composition so that things fit together harmoniously, albeit disconcertingly. Much of this would surely involve what could justifiably be called convergent thinking—critical thinking—for every visual artist knows there are clearly right and wrong positions for a component in an image, clearly right and wrong colours for a

---

* This isn't to imply he didn't have some very intelligent things to say about matters outside of physics, only that the general public treated him as a seer even when he uttered clichés.

component in the context of the other components. (There may be more than one right position or colour, but that is totally irrelevant: IQ tests may be designed to have only one right answer, but that is merely an artefact of the test, put there to make scoring easier. What matters about convergent thinking is not that there is *one* right answer, but that there are answers which are clearly right and answers which are clearly wrong.) Again I ask: Is there an IQ test—any test of congruent thinking—that, should Magritte have taken it, would've flagged his brilliance? And again I answer that, at least as far as I know, there is not.

It is anybody's guess what IQ or CQ Magritte would attain on any test of convergent *or divergent* thinking. But I'd put my money on his scores not being anything to write home to his mother about, even though his paintings are irrefutable proof of both his exceptional intelligence and his creativity—by almost any definitions. I'm no expert on Magritte's life or personality, but from what I do know, if he were incapable of following a simple mathematical proof or unable to engage in witty repartee, it wouldn't necessarily surprise me and it wouldn't matter. The only real test results of his genius are the obvious intelligence and creative wit of his work.

IQ, CQ! ADD 'EM UP AND WHAT DO YOU GET? ZERO?

So what does all this add up to?

.

First, that while the right hemisphere of our brains may be more fluent in manipulation of spatial and visual entities, and the left more fluent with verbal manipulations, there is no reason to equate the right hemisphere with creativity (and divergent thinking) and the left with intelligence (and convergent thinking).

.

Second, there is also no reason, on the basis of the predictive validity of existing psychometric tests of either convergent or divergent thinking, to trust performance on such tests as measuring what is usually meant by intelligence or creativity—or to credit these tests with the ability to predict creative accomplishment.

.

Third, creative accomplishment, even genius, probably does require some innate skill at critical and logical thinking, but such ability need not be particularly exceptional. And it is not safe to assume that conventional tests for such an ability give an accurate estimate of potential—especially in the case of creative individuals who are more inclined than most to see unexpected, divergent, answers to a question.

.

It may well be that being born with a brain hardwired to think clearly, logically and critically is one of the secret agents operating in the creative individual. And it may well be that being born with a brain that has a lot of eccentric short-circuits that result in uncommon connections between ideas may be another secret agent operating in the creative individual. But, even if we were to assume the psychological tests for these wirings have any validity, there obviously are other factors that have to be taken into account to explain creative accomplishment. The next reasonable place to look is in the even more murky area called 'personality'.

.

.

CASE STUDIES: LEONARDO DA VINCI & FRANCIS GALTON

Francis Galton, obsessed with the idea of general intelligence, and Leonardo da Vinci, a prime example of general intelligence if ever there was one—were both extreme cases of what has come to be called 'Renaissance Men'. What do their lives reveal about the relationship between creativity and intelligence?

.

Leonardo da Vinci was born in 1452 in Anchiano, near the town of Vinci. His mother's name was Catarina and little is know about her except that she was a peasant farmer's daughter. His father was Ser Piero da Vinci, a wealthy public notary. Leonardo's parents weren't married; in fact, his father married another woman on the day of Leonardo's birth. However, Leonardo was integrated into his father's family, living as a young man with Ser Piero's father in Vinci. Then when he reached his mid-teens, his father, noting his talent at drawing, apprenticed him to Andrea del Verrochio, a famous and accomplished artist working in nearby Florence.

.

As has been traditional throughout much of the history of Western art, the master artist took responsibility for the overall composition of a major work, but often delegated the actual painting of parts of it to his apprentices. It's probably apocryphal, but the story goes that Verrochio let Leonardo paint an angel in his "Baptism Of Christ", and the young man's angel was so superior to one painted by Verrochio that the 'master' resolved never to paint again.*

.

In 1477 Leonardo set out on his own. In 1482 he entered the service of the Duke of Milan, and simply abandoned his first commission: "The Adoration of the Magi". This failure to bring his projects to completion was to become a life-long habit.

.

He spent 17 years in Milan, painting and sculpting and designing court festivals for the Duke. But he also worked on designing buildings (everything from churches to fortresses), weapons (everything from cannons to catapults), and machinery (everything from flying machines to submarines). He also pursued a variety of scientific studies and produced his first anatomical drawings based on the dissection of human cadavers. His workshop was by all accounts

---

* A similar story is told about Picasso apprenticing with his father.

an insanely frenetic place, both because of his incredibly broad interests and boundless energy and because of all the apprentices and students it attracted.

·

After the French invaded Milan in 1499, Leonardo moved on, working and traveling throughout Italy. He had a number of different patrons, including the infamous Cesare Borgia. He traveled with Borgia's army as a military engineer and met Niccolo Machiavelli, author of *The Prince*, who got him a commission to paint the "Battle of Anghiari." He reputedly began work on his most famous work, the "Mona Lisa", in 1503.

·

In 1513 he again settled down, this time in Rome under the patronage of Giuliano de' Medici and also was given several projects for the Vatican. Three years later his patron died and King Francis I of France offered Leonardo a position as "Premier Painter and Engineer and Architect of the King", a position he retained until his death in 1519, at age 67.

·

Leonardo da Vinci's life was one stuffed to over-flowing with adventure and accomplishment in both the artistic and the scientific domains, and surely the current meaning of Renaissance Man as a polymath is more based on him than any other figure from that period. His passionate interests were incredibly—some would say insanely—broad in scope, and if anyone deserves to be credited with stellar intelligence *and* creativity, mastery of convergent *and* divergent thinking, it is Leonardo da Vinci. But what is of particular interest here is not his exceptionality, but rather how he exemplifies characteristics—albeit in the extreme—associated with many creative individuals.

·

To take the first, most mundane, characteristic remarked upon by those trying to connect brain function with creativity: Leonardo was an unrepentant southpaw. The claim has been made that left-handed people have right-hemispheric dominance and so are naturally more creative, more 'artistic'. But what distinguishes da Vinci from the majority of visual artists is his scientific, logical bent. He has been called the finest scientific mind of his generation, and furthermore, most of his scientific contributions (e.g., his anatomical studies) are based on meticulous—almost excessive and plodding—attention to detail. He only left a very small collection of paintings (and many of them unfinished).* None of his sculptural works was ever completed. If his stature in the pantheon of artists were based solely on the number of completed works, he would be considered a very minor figure. Indeed, the quality and innovativeness of his work is exceptional, but even that has more to do with what might be considered quasi-scientific experimentation with special effects than anything else.† The historical record of Leonardo da Vinci the man is arguably more congruent with the idea of a "left-brain" thinker than right, more stereotypical of a creative scientist than a creative artist.

However, this is not to say his brain wasn't strangely wired. He kept his journals in Latin (supposedly because he wanted to improve his Latin, feeling he had inadequate schooling in this language which was considered a mark of an educated man) and used mirror writing, not something that one would do unless it was, for some strange neurological reason, easier than it is for most of us.

The precision of his draftsmanship, evidenced very early in his life, again suggests some special neurological gift. But like many such 'gifts', it may have cost something substantial. One of the most notable features associated with his genius was his reluctance—or inability—to carry his projects through to completion. He is notorious for the number of incomplete works he left in all his fields of endeavour.

I came across an interesting web page on so-called 'mental health' where the author remarks that Leonardo da Vinci shows many of the indicators of ADD ("Attention Deficit Disorder"). There is a checklist for characteristics that flag this latest, fashionable psychiatric diagnosis. It includes the following: a personal sense of underachievement despite obvious accomplishment; difficulty getting organized; chronic procrastination or trouble getting started; many projects going simultaneously; frequent search for high stimulation; intolerance of boredom; easy distractibility; trouble with focusing attention, often coupled with an exceptional ability to, at times, focus intensely on the task at hand; trouble going through established channels or following proper procedures; impulsiveness; restlessness.

---

* He only left 17 completed paintings, which may be the smallest oeuvre of any artist considered a serious professional.
† The sfumato and chiaroscuro 'special effects' in the "Mona Lisa" are what makes it so singular. His experiments with different pigments and media are notorious because they often backfired, and so many works have not survived the ravages of time.

Examples of all these can easily be found in Leonardo's life, and, interestingly, it isn't difficult to find evidence of these characteristics in many eminent, creative people—including Francis Galton.

.

It has been said of Leonardo da Vinci that he was cursed with too many gifts: He was handsome; curious about everything, with the intellectual ability to satisfy his curiosity; possessed of a natural talent for draftsmanship and perspicacious in perception; ingenious; energetic; able to spot unusual but meaningful connections between things. Hell, apparently he could even sing!* These gifts turned him into a dilettante—albeit one of the most accomplished dilettantes in the history of Western civilization. Maybe Renaissance Man and dilettante are close to synonymous?

.

.

Francis Galton shared many characteristics with Leonardo da Vinci, although he succeeded in actually completing most of his diverse projects—perhaps because his scope was slightly more narrow (being more purely scientific) and perhaps because his brain was not quite so strangely wired. Galton was born in 1822, the youngest of seven children in a well-to-do English Quaker family. His father was a banker and his mother was the daughter of Erasmus Darwin, grandfather of Charles Darwin.

.

Galton's fascination with objectifying intelligence is particularly interesting because he himself seems to, as a young man, have displayed the dismaying syndrome so often now associated with a misguided emphasis on IQ over actual achievement—a confusing mix of arrogance and insecurity. Of course the IQ test had not yet been invented, for Galton was the one to suggest that there was such a clearly measurable, physiologically determined thing called 'intelligence'. (His own subsequent attempts to measure it were rather crude phrenological ones: he seemed to believe head size reflected brain size and brain size indicated level of intelligence.†)

.

So in Galton's case it was not the results of an IQ test that caused his head to swell: it was his parents. They were convinced—and tried to convince him—that he was brilliant, a prodigy, and intelligence personified. And it was to be expected given his lineage. The actual known facts of his childhood don't seem to support the prodigy

---

* Seriously, Leonardo was said to have had an exceptional singing voice.
† Ironically, Galton's own cranial size was actually somewhat less than average.

claim, and he probably realized this before too long. He was a mediocre student and probably took any objective failure to demonstrate his supposedly exceptional abilities as devastating proof of exceptional inadequacy. Of course, he wasn't inadequate at all. But it seems likely that he (like many young people a century later finding they had high IQs) took to heart the idea that he was exceptionally intelligent, and then had to deal with early failure to demonstrate this alleged brilliance by academic achievement.

So, despite all the expectations and self-esteem boosting by his parents, biographers have suggested that Galton's head wasn't really swollen at all. Instead, it's been argued, he may not even have been convinced of his own competence, never mind brilliance, and actually suffered from a profound inferiority complex—and that the nine books and over 200 papers he penned on such diverse subjects as fingerprinting for personal identification, applied statistics, twins, blood transfusions, the art of travel in undeveloped countries, criminality and meteorology are evidence of over-compensation for a deep feeling of inferiority after his failure in his medical studies at Cambridge. Probably he was typical in many ways of those many creative individuals who do poorly academically or feel inadequately educated—something Leonardo da Vinci also felt because of his limited formal academic education. This tendency to feel insecure about one's intellectual credentials is a recurring theme in the lives of the most eminent and accomplished individuals. It is surprising how many brilliant people will privately admit they feel like phonies, great impostors—no matter how much they accomplish.

Well, Galton did a lot—eventually. (Many of his major accomplishments date to after he suffered a three-year-long, mid-life crisis at the age of 47.) However, one gets the definite impression that Galton's own eminence, his exceptional intelligence in the sense of real achievement was more a demonstration of the tidal force of motivation and emotion than of his own theories of inherited braininess.

This is not to underestimate the importance of the genetic factors he considered so important. (He probably did inherit a damn fine set of cortical neurons.) Rather it is to suggest that this general factor of intelligence he was the first to propose, and which so many people uncritically equate with IQ or early academic success, includes more than a simple inherited talent for formal reasoning, more than a bigger brain with more neuronal connections. Genius and creativity are the result of much more than oversized brains hardwired to think logically—or, for that matter, divergently. Personality also matters. The person also matters.

Sir Francis Galton was knighted in 1909 and died in 1911. He never had any offspring.

THE SPLIT (IN) PERSONALITY

"It is by no means certain that our individual personality is the single inhabitant of these our corporeal frames... We all do things both awake and asleep which surprise us. Perhaps we have co-tenants in this house we live in."
—Oliver Wendell Holmes (*The Guardian Angel*)

"When Heraclitus spoke of the impossibility of stepping into the same river twice, he wasn't just referring to the water."
—Hippokrites

"Personality Theory", the term now used by psychologists, dates back at least to the ancient Greeks. The four basic elements according to Pythagorean thinking (air, fire, water and earth) were linked respectively to these four "humours" or bodily fluids: blood, yellow bile, phlegm and black bile. The dominance of any one of these led to personalities called (again respectively) sanguine, choleric, phlegmatic or melancholic.* A sanguine individual was amorous, happy and generous. The choleric type tended to be aggressive—even violent and vengeful. A phlegmatic fellow was cautious, rather dull, and unadventurous—even cowardly. A melancholic guy was prone to be gluttonous, lazy and sentimental. As in everything, the Ancient Greeks favoured balance and moderation: The ideal personality resulted when no one of these humours dominated but were present in roughly equal proportions.

This simple, early taxonomy is very similar *in form* to that of current 'scientific' theories of personality. Psychologists have replaced the word "humours" with the term "personality traits"—defined as an individual's consistent tendency toward, or predilection for, certain behaviours. They have applied sophisticated statistical tools such as factor analysis to try and find 'clusters' in the answers to personality test items or check lists of adjectives describing a person's personality. Gordon Allport, before the advent of computer databases and programmable searches, laboriously searched an

---

* The meaning of these adjectives in this context isn't exactly the same as in current usage, but it is close.

English language dictionary for adjectives that could be said to describe a personality trait (e.g., stubborn, warm-hearted, passionate, lazy, etc.) He came up with 17,953 words.* Subsequent efforts at defining personality have been largely based on trying to reduce this number to something more manageable by looking for inter-correlations between descriptors (or answers to test questions that putatively indicate a person possesses a particular characteristic). Traits are set up as polarities. An individual's personality is defined by how strongly he or she manifests (usually just through personal questions on a personality test, dubious as that may be) each of the traits in the particular taxonomy. Currently the magic numbers leading the competition for allegedly fundamental traits are 2, 3, 5 and 16.

.

The number five seems to have the most fans. It is called by its proponents "The Big Five Theory". So replace sanguine, choleric, phlegmatic and melancholic with extroverted/introverted, stable/neurotic, agreeable/disagreeable, conscientious/unconscientious, and open/closed to new experiences. I perfectly understand if anyone not blinded by immersion in this branch of psychology finds this almost laughably naïve and simplistic. However, given that this is the current system most often used in attempts to apply psychometric tools to describe creative people, I will have to use it as my point of reference in the following discussion.

.

So. What does the allegedly scientific research have to say about the personalities of scientists versus artists—or about creative individuals in general? What are creative folk like as persons—or personalities? Is there really a creative personality type? Is there any consistent difference in personality between creative scientists and creative artists? And what does the biographical evidence say to confirm or deny the psychometric research?

---

* Are humans self-obsessed, or what?

THE CREATIVE PERSONALITY: STRAITS, TRAITS AND ACTIONS

Mozart was cheerful and childlike, but Mahler was dour and depressed. Van Gogh was mad as a hatter, but Matisse was the epitome of sanity. Swift might be considered choleric, while Salinger phlegmatic. Probably the safest generalization one can make about personality and creativity is that diversity in the personalities of creative people is greater than in almost any other imaginable sample of humans. This is to say that virtually any statement about particular personality traits being consistently associated with creativity can easily be refuted by many counter-examples. Of course, as long as one isn't dogmatic and doesn't confuse moderate correlations with universal statements, there is some point to the exercise. There is, for example, considerable evidence that creative individuals (even actors!) are more likely to be introverted than extroverted, but there are plenty of Norman Mailers out there. In fact one of the reasons I have included 'case studies' at the end of each chapter in this book is as an antidote to the poison of broad generalizations about creativity. Sometimes more insight can be gained by looking at individual cases than doing summary statistics. Having said that, I shall now go where it might be foolish to tread.

.

The first important thing to note is that 'personality', like 'intelligence' and 'creativity', is defined in many different ways—most of which neglect a very important characteristic of the innovator in art and science. Oh yes, my teachers said it too: Genius is 5% inspiration, 95% perspiration. One might quibble with the percentages, but this particular truism is true: accomplishment of any sort requires effort, even of the most gifted. Wolfgang Gottlieb Mozart is usually portrayed as having his marvellous melodies effortlessly served up to him, while his less fortunate colleague Antonio Salieri did a lot more perspiring—to a lot less effect. However, the Köchel Mozart Catalogue lists 626 compositions created in 30 years, and that includes 41 symphonies! It may have come easy for Wolfgang, but obviously he was no slacker.

.

The idea of laziness, and by implication its opposite, was part of the classical Greek four-trait system (as a characteristic of the "melancholic"), but it is less so a part of current ideas of personality. Psychology does, of course, look at 'motivation', but somehow the idea of it being a central, defining trait is looked at only obliquely.

More often psychologists try to understand what motivates creative people to work so hard, often for so little extrinsic reward, rather than considering the centrality of motivation (whatever its source) in any understanding of creativity. The pat and singular answers, such as Freud's, as to why creators create are doomed to failure, for the aforementioned diversity applies at least as much here as to other areas. Compare Dr. Johnson's "only a fool writes for anything but money" with desk-drawer poet Emily Dickenson, certainly no fool. Methinks it fair to say that Sam and Emm had very different motives for their contributions to literature, but what they, and all creative people, do have in common is an overwhelmingly strong motivation to create, to discover, to achieve.

TWO SOURCES: TWO PROBLEMS

The human brain is not only split into two hemispheres, it seems prone to split everything into two parts—good and evil, bright or dark, art or science, yin or yang, analysis or synthesis, positive or negative. Probably two *is* a magic number, *the* magic number. And maybe the world really *is* bifurcated.

Theorists of the creative personality can also be sorted into two categories—those who view the creative personality in purely positive terms and those who view it in purely negative terms. At one extreme are theorists such as Abraham Maslow with their idea of the creative person as "self-actualized", the epitome of a well-integrated and adjusted personality, while at the other pole are those such as Sigmund Freud with their depressing idea that the creative person is a psychologically-crippled personality whose gifts are a by-product of neuroticism or even psychosis.

The relationship of creativity to madness is a theme for a later chapter. For now, I want to keep the focus on the science/art dichotomy. The positive take on the so-called creative personality is more often associated with the scientifically accomplished, and the negative with the artistically accomplished. Why this is so, and whether it is really justified, is important.

What information we have about the personality characteristics of indisputably creative people springs from two sources. These are biographical information and data gathered by the quasi-scientific studies of psychologists. From both these sources spring somewhat muddy waters.

The problem with biographical information is that it involves selective sampling—both in terms of subject chosen and features of the subject accented. Biographies are written about people who have interesting, exceptional lives. If an artist or scientist lives a very uneventful, completely conventional, even dull, life—except for his creative contributions—he or she isn't a hot candidate for biography. The great French novelist Gustav Flaubert offered this advice to would-be artists: "Live like the bourgeois!" Save one's creativity for one's art: do not squander it on life where it might interfere with one's work. And, furthermore, a biographer, like a novelist, selectively samples for the good, juicy bits from available material, so as to make the tale of a life interesting and engaging. For example, most writers spend most of their time alone in their rooms working. Well that doesn't make for riveting reading: Jane got up, wrote for four hours, had grilled cheese for lunch, did some correspondence, revised a few proofs, had dinner, and then read a book until bedtime.

.

If you travel and you have an uneventful journey, with no missed trains or mishaps of any sort, you aren't going to be able to write a good travelogue. It may be that artists travelling through life are more prone to adventures (which just might be another word for mishaps) than are scientists, and so their lives are more likely to be the subject of biographical report. And it may be that this is so because artists have more need of life experience as raw material than do scientists, but then maybe it is merely that they are less rewarded for their innovations—and thus suffer more from the slings and arrows of outrageous fortune—or lack of fortune. This is not to say that there haven't been eccentric scientists whose sometimes-tragic lives make interesting biographies, only that overall external factors may—as pollsters put it—bias the sample.

.

The other source, psychological—allegedly 'scientific'—studies so often are based on highly questionable assumptions and theoretical systems that this source too has to be taken with more than a few grains of salt. There may be something to be said for some of the research, but so much is so contrary to what creative people say of themselves, never mind common sense, that those outside of the inbred psychological enclave have a very hard time taking it seriously. Again, I'll put off to a later chapter a more detailed discussion of the so-called science of psychology as it relates to the issue of creativity, but for now I'll have to treat it with some credibility in the context of differences between the scientific and atistic types.

.

Before trying to find these differences, it is worth looking for commonalities. As already mentioned, the currently most popular number for 'fundamental' personality traits is five: extroverted/introverted, stable/neurotic, agreeable/disagreeable, conscientious/unconscientious, and open/closed to new experiences. Demonstrably creative people have been charmed, coerced or bribed into taking personality tests, and some of the results are in.

.

Now, remember these are summary statistics! (Summary executions?)

.

Creative people are introverted, conscientious, open to new experiences, disagreeable, and neurotic.

Before proceeding, some explanation is needed as to how this information was gathered and what is meant by, for example, "introversion" or "neuroticism". Personality testing is often surrounded by an aura of mystery which when uncloaked seems so simple as to be silly. Consequently, to understand what these findings really mean, it is necessary to digress slightly and examine the nature of the tests on which they are based.

SORRY, YOU FAILED YOUR PERSONALITY TEST!

Say I want to invent a test to measure extroversion/introversion and neuroticism/stability. I make up a number of questions that on the surface seem to relate to these traits.* I might ask if you'd prefer to stay at home and read a book or go to a party. If you opt for the party, you get a point for extroversion. Psychometricians would say the question "loads" on the extroversion scale. I might then come up with a question that asks if you easily become nervous and tense. If you answer in the affirmative, you just 'scored' a point on neuroticism. I give this test to a sample of people and tabulate their scores on the two scales. These tabulations are called *norms*. Then I give you the test. Your results won't simply state something like "you *are* extroverted and neurotic". They'll say something such as "you scored at the 87th percentile on Extroversion and at the 72nd percentile on Neuroticism." This simply means that you answered more questions intended to measure extroversion with the 'extroverted answer' than 87 out of 100 people previously tested to develop the 'norms', and you answered more questions loading on the neuroticism scale with the 'neurotic answer' than 72 out of 100 people from the normative sample. Now of course it is technically legitimate, but certainly not very meaningful, to thus say, "Hey, you are extroverted and neurotic."

.

From this description of what really is involved in personality testing, the problems are fairly obvious, but, nevertheless, I'll quickly review them.

.

First, there is the convergence problem associated with there being only one 'right' answer; i.e., an answer either does or does not load on the scale for a particular trait. Ask me, ask most people, if they'd prefer to stay home and read a book or go to a party, and they will reply with the questions: "Whose party? What book?" Depends. I like to read and I like to party. But I show some discrimination in these matters. If invited to a party where most to be in attendance were accountants, I'd almost certainly pass. But if the book in question were a textbook on double-entry bookkeeping—well even that party would seem a better use of my time. Many people are justifiably annoyed with having to pick a single answer to such questions, and it isn't unreasonable to assume that they often pick an

---

* This is sometimes called "face validity".

answer almost at random. Furthermore, it seems to follow that if creative people are more inclined to divergent thinking, they are ones for whom the answers chosen are even less likely than usual to be really measuring what the test designer intended.

.

Second, there is the problem associated with the percentiles and their simplification. The hypothetical 72nd percentile score you received on neuroticism is above the average of the normative sample, but only slightly. Even if this could be unequivocally taken as a valid measure of neuroticism, it would be misleading to call you neurotic. This problem is further confounded by grouping the results of a highly diverse sample of individuals, such as, say, painters. It's bound to be a bit misleading to average the scores of the demonstrably unbalanced Van Goghs and Pollocks with the more emotionally stable Monets and Matisses. Proper understanding of the results requires more statistical sophistication than is common—or in many people's minds, worth the effort.

.

Third, and most importantly, there are the problems of reliability and predictive validity. These are complex issues in psychological testing that are not easily summarized. What matters in the present context is how they relate to the findings about creative people. Are the results of recent personality tests on accomplished artists and scientists reliable in the sense of being likely to match up, correlate, with the results of testing creative individuals a generation or a century from now? This is, of course, an unanswerable question, but there is reason to be doubtful—by looking backward and extrapolating forward. For example, artistic endeavour in the Middle Ages was more often a collaborative effort than now, so one might assume the artists creating the great cathedrals of Europe were less introverted than the typical visual artist of today working alone in his studio. And more and more science is being done as a collaborative, team effort than it was even in Einstein's time, so again the extroversion scale scores of creative scientists might be assumed to have increased this century. (These, although superficially sensible, are just speculations, of course, and might be wrong.)

.

Regarding the issue of predictive validity, one has to ask exactly what particular test results on a practising scientist or artist *should* predict. If poets score lower than average on agreeableness and higher than average on neuroticism, does this predict more failed marriages? (Watch out, he may write you beautiful love poems, but he'll be a pain in the ass to live with!) In short what do we really learn about creativity by personality test results, even if one credits them with some commonsensical validity? One thing is certain: you cannot flip the equation. If a person tests as introverted, neurotic, disagreeable, conscientious and open to new experiences, it doesn't mean one can predict that person will be creative. If you test thousands of mathematicians with the question "Are you male?" you will find that most answer in the affirmative. Obviously that doesn't mean that you can assume that the next man you meet will be a mathematician, nor that any woman you meet will not be!

.

Finally it should be mentioned, as an aside, that the common misconception that personality tests will tell a person something about himself he didn't already know is not only illogical but downright ridiculous. You know more about yourself than anyone could possibly determine by asking you a set of standard questions. If you didn't know the answers to the questions already, how on earth could you answer them? What you *might* learn is how a select sample of other people has usually responded to the questions. If your extroversion score comes back as at the 95th percentile, you've learned something about what percentage of people in some sample answered certain questions about sociability the same way you did. But you haven't been given any deep insight into a hidden aspect of your personality: surely you must already know you're a party animal.

THE BIG FIVE AND THE BIG QUESTION

Given all these qualifications, the big question remains: What does it mean to say creative people are more often more introverted, neurotic, disagreeable, conscientious and open to new experiences than average?

Introversion. The term is an old one meaning inward looking, but it has only come into common usage (a somewhat different usage) since being adopted by psychologists, who seem to primarily define it in terms of opposition to extroversion. The tendency is to think of extroversion as sociability, so introversion has come to imply unsociability. There is also a tendency to think of the extrovert as more energetic and cheerful. Actual test questions for extroversion/introversion to a great extent, although not entirely, reflect these meanings. The extrovert prefers the company of others; the introvert prefers his own. The extrovert looks out to the world of others and defines himself by what is reflected back to him. The introvert looks inward and generalizes from what he finds there to the world outside.

There is nothing particularly surprising about finding that artists and scientists are more introverted, if one conjures up the image of writers toiling in the privacy of their studies or scientists burning the midnight oil in their labs. However, even actors seem to average below normal for extroversion.

One obvious reason creative people might be less sociable is that they need to be alone to get their work done, and because getting their work done is important to them—the one indisputable fact about creative people. Naturally they value their private time. This is even true of those in the performance arts or scientists working together on a project.

Furthermore, introversion and introspection have the same prefix. One has to look inward to discover relationships, be they highly personal, as may often be the case with writers, or impersonal and abstract as is the case with scientists. In short, one has to get far from the madding crowd to find the peace to *think and reflect.*

Conscientious. Again there is nothing surprising about this trait being associated with creativity. But it should be noted that the conscientiousness might not apply to every aspect of the person's life: it must, however, apply to the creative endeavour. Rembrandt, typically, was far from conscientious in his personal affairs. After his death, old uncashed cheques, payments for his work, were discovered. He simply was too busy with—too conscientious about—his work to bother to collect money due him, even though he was quite poor.

.

As already remarked, being highly motivated is more often considered by psychologists to be the result of current and temporary external factors than it is to be seen as a trait, as a relatively stable feature of a person's character. This makes understanding the perseverance of the starving artist inexplicable. I think that hiding under the label "conscientiousness" is the stable trait of being highly motivated to do, and do right, what one values. Without this, no amount of talent would carry an individual through to the end of writing a novel or through years of work hoping to arrive at the solution to some complex mathematical or scientific problem.

.

Open to experience. Some people like to be in a rut. Routine makes life easier. If you're comfortable, why should you risk discomfort? Resistance to change is a stronger motive in most people (at least those in comfortable circumstances) than is the drive for novelty. But new ideas don't sprout in ruts. Because innovation, new ideas, new output can only be generated by new input, only those people with an exceptional willingness to have new experiences will ever produce anything new. Of course, these new experiences can be found in books or correspondence or conversation. It is said that the philosopher Immanuel Kant led such a regulated, bourgeois existence that people set their clocks by when he passed their houses on his daily walks. However, the truth is he was quite gregarious until his later years, and he always remained open to new ideas. Openness to experience doesn't only mean traveling to foreign places or experimenting with unusual foods or sexual practices.

.

As mentioned in the profile of Da Vinci, he had at least two characteristics that are on the checklist for diagnosing Attention Deficit Disorder—frequent need for stimulation and intolerance of boredom. In reviewing the lives of great artists and scientists, this seems to be a fairly common thing. And while it may lead, only too often, to experiences that are uncomfortable or even personally destructive, it also leads to new insights and understanding.

.

Disagreeable. The meaning here is not disagreeable in the sense of 'unpleasant': it is much more literal. Creative individuals do not agree with much of the current or popular 'wisdom' that surrounds them. Of course if they insist on always, relentlessly, expressing this disagreement, they can be very annoying to those around them. But the trait refers specifically to the tendency to not automatically, unthinkingly agree with currently accepted ideas or other people's judgments. In other words, this trait could be more positively expressed as independence of thought. It is self-evident that this is a prerequisite to innovation. The innovative visual artist must see the world in a new way: The Impressionists had to look at the world as a play of light, not a formal composition of objects stamped with approval by academic critics. The scientist that causes a genuine paradigm shift has to disagree with whatever is the currently dominant theory: Galileo had to reject the Ptolemaic theory of astronomy—and he annoyed some very powerful people in doing so.

This closely relates to scepticism, what might be considered *the* defining characteristic of truly scientific thinking. Since most people seem to tend more to gullibility than scepticism*, the outspoken sceptic is annoying. No one likes to have their beliefs questioned or mocked, no matter how superficially those beliefs were acquired. It is amazing the amount of hostility an outspoken sceptic (of even such patently absurd notions as divining for water or clairvoyance or homeopathy) can generate. For example one such famous sceptic, James Randi, a basically amiable stage magician turned debunker of charlatans and paranormal balderdash, is vehemently hated by many, many 'agreeable' folk.

Scepticism among artists, as compared to scientists, is less obvious. In fact many artists actually seem exceptionally gullible, often unquestioningly accepting the most outrageous ideas. But the scepticism is usually still there: it is just focused on different areas, such as the validity of current social mores or dominant aesthetic values. The apparent lack of scepticism among artists toward flaky pseudo-science is, at least partially, because many artists today still only know the terrain on one side of the Two Cultures Chasm.

It must be remembered that scepticism, or disagreeableness—or for that matter, any trait—is not going to be manifested in every aspect

---

* One poll of the American public allegedly found that about 80% of the sample believed there was some truth to astrology.

of a person's behaviour. One can, for example, thoughtlessly agree with abhorrent doctrines such as Nazism or the moral legitimacy of slavery and still be sceptical of questionable scientific ideas or ingrained, outmoded aesthetic values.

.

On the other hand it is also true that a tendency to question, a tendency to think independently, a tendency to be sceptical—is unlikely to be limited to one domain only. Surely this is one reason that creative individuals are so often labelled "non-conformists" and make many conventional people uncomfortable.

.

Neuroticism.

.

I've saved the worst for last. Lip service is paid to the official doctrine that mental illness is no different than physical illness: they aren't value judgements; they're just statements of fact. But the truth is that there is a world of difference between saying "He's diabetic!" and saying "He's neurotic!" When most people say someone is nuts it isn't a value-neutral statement.

.

The relationship of mental illness to creativity is a topic for more detailed examination in a later chapter, so I'm not going to say much about this fifth characteristic—except to try and clarify its definition in the context of psychological tests. Neuroticism as a trait measured by personality inventories does not have the same meaning or connotations as it does in popular usage. If one examines the questions on a personality test that load on the neuroticism scale this becomes evident. If you admit to being tense in social situations or having trouble sleeping or worrying about your health, or sometimes feeling your heart racing—well then you rack up points on the 'N' scale. An expression that more clearly describes what such questions measure is the colloquial phrase "high strung".

.

Often people are high strung because they are full of energy, highly stimulated. So? Coffee is a stimulant. It has been shown to heighten alertness, improve judgment, speed reaction time, and even enhance cognitive functioning. It also speeds up heart rate, makes one edgy and tense, and gives many people insomnia.

.

Without dismissing higher scores on neuroticism as meaningless in terms of understanding the relationship of emotional stability and creativity, it is worthwhile to consider how such scores might naturally follow from most of the other aforementioned traits. If one is open to experience, one is inevitably going to be put in situations that induce tension. A person in disagreement with prevailing attitudes and beliefs is going to be under constant stress and understandably tense. And it is always the conscientious one, not the laid-back fella, who feels pressured.

COMPARE AND CONTRAST: NERD AND FLAKE

So much for gross generalizations about creative artists and scientists—now on to the differences between them. What are these differences?

.

One way of approaching the personality differences between artists and scientists is to consider what is *required* of them to be successful in what they do. It is not always true that aptitude and interest and personality coincide (and it can be tragic when they do not), but usually they do. If you are good at something, usually you are drawn to it. If you discover you have a knack for word manipulation or figuring out how things work, it is only natural to find writing or science appealing. Skill is its own reward. And if one's personality inclines one to certain situations or activities, it is likely that the area of creative endeavour one chooses will match up.

.

Introversion. The artist, more than the scientist, needs time alone, works alone, and so, although both are more introverted than average, the artist is the most introverted. And within the arts, the arts most requiring time alone—such as writing—have the practitioners that are least extroverted.

.

Incidentally, writers, as a group, are a good example of the difficulty of evaluating personality. They are often perceived as being extroverted, despite the solitary nature of their work. Their lives, because of the imbroglios they often make of them, have great biographical potential. But the fact that a person's interactions with others, the love affairs and public misbehaviours, are flamboyant and contrary to conventional mores is a poor yardstick by which to measure introversion. Again, we have a problem with selective sampling. If I may be indulged, I'll use myself as an example: most of my friends and acquaintances and students think I'm extremely extroverted. I'm not. I know this because I know myself.* I spend the vast majority of my time alone in my study. And I don't just do this reluctantly. I like being alone. I become extremely cranky if, as sometimes happens when I'm travelling with family, I can't get my daily fix of solitude. But the only time other people see me is, obviously, when I'm not alone. Since I spend such relatively little

---

* And, as should be obvious, while I don't think my scores on a personality test particularly meaningful, they do agree with this self-evaluation.

time interacting with others, I probably 'over-interact'—and apparently give the impression to many people of being very sociable.*

.

With scientists the exchange of ideas and findings is so necessary and rewarding that extreme introversion is less likely. It may or may not be true that the era of the lone scientist is forever over, but it is true that more and more science has become a collaborative effort—at least experimental science. This is partially because of the cost of equipment required as science probes deeper and deeper into Nature. You can't just scrounge together a few bucks to buy an elementary-particle accelerator for your basement lab. And as funding becomes more and more important, the bureaucratizing of science is almost inevitable, with the associated emphasis on 'team-effort'. Purely theoretical science remains a relatively small part of the enterprise of science—and even it is dependent on experimental results. Einstein sitting alone doing a *Gedankenexperiment* (thought experiment) is not typical of contemporary science. Even in math, where the equipment (pencil, paper, chalk, and blackboard) doesn't cost much, social interaction is surprisingly important. More and more even math has become a collaborative effort. One of the great mathematicians of our time, Paul Erdös, is a typical, albeit extreme, example of how most mathematicians work these days. This phenomenally prolific Hungarian mathematician was peripatetic to the extreme—not really having a home, but instead travelling around the world from conference to conference, math department to math department, relying entirely for his personal survival on the hospitality of colleagues. And collaborating with so many of them that every mathematician now knows of what are called Erdös Numbers. Erdös' own number is 0. Those who have written a joint paper with him have an Erdös Number of 1. Those who have written a paper with an Erdös collaborator, but not Erdös himself, are given a number of 2. And so on. To grasp the extent of collaboration in contemporary mathematics consider that 507 mathematicians have an Erdös Number 1 and 5897 have a 2.

.

Conscientious. Although there seems to be some evidence that the scientist is slightly more conscientious than the artist, as defined by scores on conventional personality tests, this may have more to do with the phrasing of the questions (and the more scientifically oriented biases of the questionnaire developers) than with any real

---

* Or some less complimentary term.

differences in conscientiousness *toward work*. The biographical evidence is overwhelming that those working in the arts live more disordered lives than those working in the sciences, and that a disorderly life is often the result of a lack of conscientiousness in handling such quotidian concerns as finances or family responsibilities. This does not mean the writer or artist isn't extremely conscientious regarding what really matters to him—the *work*.

.

In fact, perfectionism about their work is an attribute often found among the creative in both science and art. The psychological term for a neurotic disorder that has some relevancy here is "obsessive-compulsive". The obsessive-compulsive is one who has to be sure everything is right, everything is done right, and everything is in its place. The commonplace, extreme stereotype is the person who compulsively washes his hands fifty times a day, and never leaves the house without checking that all the lights are off, the dishes done, and all the windows locked. He is also the writer who, before word-processors, would retype a whole story just because he wanted to change one character's name. Perfectionists are obsessive-compulsive—at least in regard to whatever they are perfectionists about.

.

Now if one extends conscientiousness to the extreme, you have perfectionism and obsessive-compulsive behaviour. In certain contexts (such as scientific research or artistic creation) this is not a bad thing. I like my airline pilots to be perfectionists, and I sincerely hope they are obsessive-compulsive about every dial and gauge being triple-checked before they fling me 3,000 feet up into the air in a big hunk of metal. Similarly, I think it a good thing if the scientist testing for potential side effects of a new drug is extremely careful that his test-tubes are clean and his statistical analyses repeatedly checked. Less obvious, and of less drastic consequence, is the importance of this in the creation of a work of art. But I don't think there is any doubt that the writer who rewrites and rewrites and rewrites until he gets his work as close to his idea of perfection as he possibly can is the writer who I want to read—and usually is the writer who creates the most enduring literature.

.

Open to experience. The differences here may not be quantitative; instead they are primarily in terms of the kinds of experience the person is open to. Artists, who derive so much of their working materials from daily experiences and social experiences and the emotions these experiences elicit are probably more willing to experiment with drugs and form unusual personal relationships than are scientists.

Scientists need not be so adventuresome personally or socially, for they can be good scientists just as long as they are open to new intellectual experiences, to new ideas. Still, as already remarked, personality predilections are usually not restricted to one domain of a person's life, so scientists are still much more open to experiences of all kinds than is the average person. Darwin was no homebody, Madame Curie no social conformist.

Disagreeable. The differences here don't seem significant, but scientists do seem to be more agreeable in the sense of 'fitting in'. Although typically neither suffers fools gladly, the creative scientist's disagreeableness is often more narrowly focused. Many scientists are shockingly nasty to those maintaining rival theories, but they usually have little conflict with their neighbours in the suburbs. Artists, on the other hand, are more often questioning societal values, living on the margins of society, and engaged in activities that the average person considers either meaningless or offensive. If Dr. Brown is doing research on the evolutionary basis of blood disease, the fact that his neighbour Mr. Jones is a Jehovah's Witness isn't likely to inspire any disagreeableness as they discuss the weather over the back fence. But the writer who publishes a satire of religious fundamentalism wouldn't want to live next to Mr. Jones. What the artist so often does, at least in the last hundred years, is publicly disagree with the widely held beliefs of his contemporaries, and so is much more likely to be in conflict with more people than is his counterpart in the sciences.

Neuroticism. This is the trait where the difference between the scientist and the artist is greatest. The relationship of mental instability is the subject of a whole chapter, but this is the place to note that there is little doubt that this is a greater problem among those working in the arts than those in the sciences.

No doubt part of the explanation for there being less neuroticism among scientists is the greater societal acceptance and respect they receive. Usually concurrent with this is greater financial security. Nonetheless, the extent of neuroticism among artists can't be entirely attributed to the indisputable fact that poverty and lack of respect can make one crazy—or at very least "high strung". Insofar as we are born with a greater or lesser predilection to be tense and nervous, it seems the worker in the arts does differ from the worker in the sciences.

CASE STUDIES: OTTO RANK & KENNETH REXROTH

Creativity and the creative personality have not been major concerns of psychologists. This may seem strange, given the importance of the topic and the expectation that psychology would attract people interested in the humanities. Furthermore, one would expect that Freud's interest in literature and writers would've had a more long-lasting and pervasive influence than it has. Irving A. Taylor in his introduction to *Perspectives In Psychology* presents a "Retrospective View Of Creativity Investigation" and explains psychology's lack of interest in creativity by its youth and wannabe status as a science: "In modeling their science after the physical sciences, psychologists have generally devoted their attention to relatively less complex modes of behaviour such as sensation, perception, motivation, and learning." Taylor goes on to say that, at the time of writing (1975), things were changing. And well they may have, but not that much—or not in a particularly meaningful way. It may just be that psychology really is most useful and closest to being a real science when it concentrates on more physiological matters such as sensation and perception, and on areas more amenable to controlled experimentation such as learning and memory. But the limits of psychology are the topic of another chapter.

For now, here is a brief look at psychological attempts to understand creativity from two perspectives—that of an investigator and that of a subject of investigation: Otto Rank was a psychoanalyst whose primary interest came to be creative artists, especially writers. Kenneth Rexroth is a poet who volunteered to be studied by the IPAR* group investigating creativity in the 50's and 60's.

Otto Rank was born in Vienna in 1884, son of an artisan jeweller and his wife. His older brother studied law, but Otto became a locksmith because his family couldn't afford higher education for both boys. He was alienated both from his father, who was an alcoholic; and from the dominant Viennese culture, for the Ranks were Jewish and Vienna was predominantly Catholic.

He showed an early and deep interest in philosophy, literature and music. Before he was 21 he'd read Freud's *Interpretation Of Dreams* and, inspired by it, went on to compose an essay on "The Artist"

---

* Institute of Personality Assessment & Research.

explained in psychoanalytic terms. Rank's physician was Alfred Adler, part of Freud's inner circle; and so the essay was passed on to The Father Of Psychoanalysis, who, presumably flattered and impressed, promptly took Rank under his wing, appointing him as secretary of the Vienna Psychoanalytic Society in 1906 and supplying him with the financial and moral support to obtain a Ph.D. from the University of Vienna in 1912—the first to do so with a psychoanalytical thesis topic. Rank was for a while The Psychoanalytic Society's resident expert on the arts and artists.

.

Eventually Rank, like most of Freud's disciples, fell out of favour with The Master, but not before he had established himself as a "lay" psychoanalyst* and ingratiated himself with the artistic community. His theory of the artist was more flattering than most, perhaps because he dabbled in poetry and so fancied himself an artist. He certainly held artists in high esteem. For him the artist was the successful one in coming to terms with all the problems of birth† and childhood. Rank placed a strong emphasis on will. (He could be considered an existentialist because of his belief in self-determination of self—in strong contrast with Freud's fundamentally deterministic worldview.)

.

According to Rank, there were three ways to deal with the parental and Oedipal problems of childhood. One could give in, conform, and identify with one's parents. This gave one emotional stability, but at a great cost to one's personal fulfillment. One could break away only partially, and this path led to neurosis and guilt. Or one could take the artist's—the creative individual's—path and make a complete break, pursuing one's own vision without regret. It's not surprising he was popular with the artistic community.

.

Rank's own life bore more resemblance to that of an artist than that of a scientist—at least in terms of stability. He was an outsider from the start. His personal life was turbulent and the source of his inspiration, especially after 1926, when Rank moved with his wife from Vienna to Paris. There he became involved with the literary expatriate community that included Henry Miller and Anais Nin. He became emotionally and sexually involved with Nin—not entirely a

---

* No pun intended, despite his acknowledged affair with that superb writer, Anais Nin. In this context, 'lay' simply means he was not a medical doctor. To be officially designated a psychoanalyst these days, one has to hold a medical degree—although only God and the Psychoanalytic Establishment knows why.
† The so-called "birth trauma" figured in Rank's early psychoanalytic musings.

prudent thing to do, given that she was to become famous for her candid, literary diaries. He emigrated to the United States in 1935, and four years later Knopf published his major work on creativity, *Art and Artist*, a book largely informed by his bohemian days in Paris. The same year he divorced and remarried. Further personal turmoil resulted from a rift with Freud, followed by The American Psychoanalytical Society expelling him for alleged "emotional instability" and suggesting his former patients undergo "re-analysis". Rank, however, continued to practise his personal brand of psychoanalysis and to teach at the University of Pennsylvania. His intention was to move to California—an environment at the time probably well suited to his 'artistic temperament'—but he suddenly took ill and died at the age of 55.

Otto Rank's life and writings are instructive, most notably in terms of the relationship of his views with his own aspirations and pretensions. Those psychologists envious or suspicious of artists are more likely to analyze artistic creativity in negative terms. Rank, on the other hand, was sympathetic to writers and artists, either because he counted himself as one of them or else just out of genuine admiration for their creations. So his take on them is clearly biased in a positive direction. It is hard to credit with objectivity his notion of the artist as self-fulfilling hero, especially given his experiences with the crazy and oft-times sordid lives of Nin and Miller and other assorted artist expats in Paris. His three path analysis may have some truth in that those who unquestioningly conform to parental values and expectations find life easier, and those that partially revolt are going to have emotional conflicts, while those that can really follow their own calling without regret or guilt are the most sane, successful and well-adjusted. Unfortunately, the idea that artists fall into this last category isn't supported by even the most cursory investigation of their lives.

In fact a surprising number of artists feel so poorly adjusted that they have turned to psychologists and psychoanalysts for help. Some, of course, make a conscious decision not to seek the insight into themselves that these people claim to offer because they fear such self-awareness will damage their creativity; because they fear that the cost of sanity will be their ability to paint or write. Nevertheless, in both cases these artists are assuming that psychologists really can offer deep insight into their personality and creativity.

But not all of them think so. Consider Kenneth Rexroth, a brilliant writer of many talents with an unfettered disagreeableness and scepticism toward conventional 'wisdom'. He was particularly fond of taking pot shots at the sacred cows in academic farms. And his aim was damn good.

.

Kenneth Rexroth was born in 1905 in South Bend, Indiana. He moved to California in the twenties, settling in San Francisco in 1927. He was an autodidact and polymath and a very successful one—with a breadth and scope of knowledge that seems superhuman.

He published about 25 books of poetry and plays, 13 books of translations (from Greek, Spanish, French, Japanese and Chinese), and 8 collections of essays on topics as diverse as literature, ecology, philosophy and world religion. He started the "poetry renaissance" movement in San Francisco that, in turn, gave birth to the 'school' of Beat Poetry and the fame of Allen Ginsberg, Gary Snyder, Michael McClure, Philip Whalen, and others. He had a weekly book review program on KPFA radio in San Francisco, wrote a series of columns for *San Francisco Magazine*, and penned a series of essays for *The Saturday Review* called "Classics Revisited" which eventually evolved into two books containing over a hundred insightful re-evaluations of monumental works of world literature. He was a music critic with a profound understanding of jazz. Nor were the visual arts outside his range: he was a talented painter. Somehow he also found time to get involved in social activism and the labour movement.

.

In the sixties he taught briefly at San Francisco State University and the University of California at Santa Barbara, but he was not attracted to Academe. In fact, academics with their typically narrow and specialized approach to matters intellectual and artistic were annoying to him and often the victims of his acerbic scepticism toward 'The Establishment'.

.

So the researchers at IPAR should have known they were asking for trouble and playing with fire when they asked Rexroth to be a subject in their study of creative individuals. Rexroth agreed, probably because of his 'openness to experience'—and because it would give *him* a chance to study *them*. One can imagine him thinking: "What are these silly psychologists doing anyway?" Also, as he admits, there was the fact he was to be paid to be a guinea pig.

.

The result of his experience is a charmingly snide essay called "My Head Gets Tooken Apart".* It begins: "Lawrence Lipton has a very funny poem called 'I Was a Poet for the FBI. I have never precisely been a poet for them, though I doubt not but what I have caused them sufficient annoy.† However, I now feel I have really made it. Fame has arrived. I have been a poet for the IPAR, formerly the OS of the OSS…an outfit calling itself the Institute of Personality Assessment and Research, which had started out in life as the Office of Selection of the OSS‡."

With typical irreverence he describes the experience of being a "creative person" assessed. "Everybody was so nice. I met people I knew, looking sheepish, and Jungians looking like séance addicts about to get a message from the Beyond, and clinical psychologists looking like Buick dealers, and psychiatrists fresh from the loony ward looking tired and worried, and professors looking like professors, and specialists in the creative personality (believe me, it is possible to make a living at this speciality) with the shy, happy, chummy look of specialists in the creative personality, and a couple of rugged sharpies who looked like they were in it for kicks, and several earnest and not very happy young women."

One can probably imagine how he describes the Rorschach Test, where one says what one 'sees' in inkblots, or the Thematic Apperception Test, where one makes up stories for corny pictures. Reading his description of the assessment procedures one gets a very different perspective than from reading the putatively scientific papers this IPAR study generated.

I can't resist one more quotation describing his assessment. "I made up a picture out of little bits of color. I had a long chummy, deep-speaking-to-deep, sort of talk with the expert on the creative personality. I indicated my preferences in Scotch tartans. I chased the elusive gestalts through lines of bric-a-brac. I sought the still more elusive after-images of melodies like *Humoresque* and noises like something falling down. I sorted things and interpreted symbols. A

---

* This essay can be found his book *Bird In The Bush* published by New Directions in 1959.
† This is a reference to Rexroth's unabashed association with socialism, radicalism and the labour movement, something the FBI certainly viewed with great suspicion in the late fifties, the McCarthy era.
‡ Office of Strategic Services. This intelligence organization, formed during World War II, was the precursor of the CIA.

rather frightened, puzzled, but very determined looking young woman took me in the attic, blindfolded me, led me into a dark room, and spent twenty minutes finding out if I could tell vertical from horizontal. Honest to God, cross my heart, hope to die. I could, pretty good."

.

It should be obvious that Rexroth didn't take all this too seriously, and reading it is probably a good antidote for taking any psychological research on creativity too seriously. But he does end the piece on a serious note: he criticizes the waste of money on what he considers dubious 'research' when there are more pressing concerns. Of psychologists he says the "serious ones are far too busy in clinics and hospitals trying to help the really mentally ill to bother with nonsense like this."

.

In response to this criticism (Rexroth's essay was published in a national periodical), the chief researcher responded (rather graciously and good-naturedly, it must be said) in his introduction to a big book on the IPAR project. He made the valid point that attempts to understand creativity, just as attempts to understand anything, have inherent value even if the methodology used is still primitive. One could add to this rebuttal the question: if Rexroth really felt this was a total waste of time and money, why did he agree to participate? In the essay, he admits to two reasons—"fun" and getting paid for it. This hardly puts him on high moral ground.

.

Nevertheless, even if his wasting-money criticism is flawed, "My Head Gets Tooken Apart" is well worth reading for his other criticisms and clear-headed comments about what does and does not make sense in the psychological research on creativity. He turns the tables on the researchers. In this trenchant essay, it is their heads that get "tooken" apart.

.

Kenneth Rexroth died on June 6, 1982 in Montecito, California, and because of his deliberate estrangement from academics and the literary establishment is not nearly as well known or as frequently read as he deserves.

THE SOLEMN FRIVOLITY OF ART AND
CHARMING FRIGIDITY OF SCIENCE

"Art? What is art? Art is a man's name"

—Andy Warhol (attributed)

"The scientist does not study nature because it is useful; he studies it because he delights in it, and he delights in it because it is beautiful. If nature were not beautiful, it would not be worth knowing, and if nature were not worth knowing, life would not be worth living."

—Jules Henri Poincaré

You'll meet him at a cocktail party for Professional Educators, the fellow who maintains that "everyone is creative", the cheerful fellow who thinks creativity in art and science are not really different. He might have unnice things to say about C.P. Snow (if he's heard of him) because he'll think Snow was in favour of the division between art and science! He won't know very much about art and probably even less about science, but he'll profess great respect for both. He'll almost certainly talk about the educator's* responsibility to encourage creativity in young people and probably have a few unkind words about evaluative testing and having a standard curriculum. He might even prefer the label "facilitator" to "educator". When you ask if he facilitates in a Faculty of Education and whether he's heard the one about those who can't do, teach, while those who can't teach, teach teachers, his friendly smile will only fade ever so slightly. He's not a disagreeable sort of guy. So it may seem strange that a lot of artists and scientists find him as offensive as flatulence in the kitchen.

I think this has to do with the arrogance born of ignorance that Professional Educators unconsciously display when they make pronouncements about that of which they know naught. Art and science are very different activities and to understand those ways in which they are similar requires some sincere, empathetic understanding of each. And not everyone is creative—at least in any meaningful sense of the word. A man or woman who has had the talent and perseverance to write a symphony or formulate a complex

---

* AKA "teacher".

scientific theory isn't charmed to hear some overpaid facile 'facilitator' implicitly lay claim to an equal measure of creativity and accomplishment.*

.

The last chapter looked at artists and scientists, the creators. This chapter looks at the nature of the act of creating art and doing science.

---

* This is not to say that teaching isn't an art, with the socially worthwhile product being an educated person. It is to say that one should know of what one teaches, and this is less and less true as education becomes more and more 'professionalized'. Practicing artists and scientists who have never taught do not understand the art of teaching—the skill, effort, and talent required to do it well. 'Facilitators of creativity' who have never really done science or art should not be babbling on about how everyone is creative.

ART: THE SERIOUS NATURE OF PLAY

You look out your front window. Two young boys are standing on your lawn. They both have their fists cocked. Suddenly one of them ducks down and charges the other, knocking him to the ground. You rush out your front door to find them wrestling each other, rolling dangerously close to your carefully tended flowerbeds. You dash over and pull them apart—partly out of a sense of adult responsibility to prevent children from harm and partly (let's be honest) to prevent them from crushing your beloved dahlias. And how do they respond? In unison: "Hey, mister, relax, we're just playing!"

Of course their combat looked serious. That is because play *is* serious business for children. In fact play seems to be serious and essential in all organisms high enough up the evolutionary scale for us to recognize it. A major part of the charm of puppies and kittens and all manner of young animals is their playfulness.

What are the young doing when they play? They are rehearsing. Whether we speak of kittens 'hunting' a ball of yarn or those young boys on the lawn, what they are doing is rehearsing for the role of adult. This is directly useful for many species, but kittens and boys are special cases. The kitten will never need to be a hunter: his food will come in a can. The boys will never need to physically battle for dominance and survival—assuming your lawn is a suburban one, not an inner-city patch of weeds. So does the kitten *need* this rehearsal? Unlike his cousins, the wild cats, probably not. Does the boys' play-acting serve any purpose? Yes, it still does. But not because of the practice it gives at combat, but because it exercises the imagination, develops the ability (perhaps unique to *Homo sapiens*) of hypothetical thinking.

Children make up stories and act them out. They imagine scenes and draw them. Some even invent melodies. Some kids are very good at this and some are not so good. (When a gang of kids get together to play act an adventure, they all know which of one of them to rely on to supply the plot.) Some kids derive so much pleasure from exercising their imaginations that they never give it up. And some never give it up just because they find so little pleasure in the real world. Most, of course, 'grow up', put away childish things, and concentrate on getting that MBA or a well paying job at the local plant.

Because so many follow the latter path, and because the exercising of imagination is so obviously a part of childhood, a number of dubious conclusions about creativity have been widely accepted. It is, for example, a cliché with great currency that all children are naturally creative until the school system destroys this natural, innate creativity: Kids are told to colour inside the lines and no, a green face is not acceptable. Well, always willing to think the worst of the school system, my knee-jerk response is agreement, and the biographical evidence we have of unquestionably creative people offers indirect support for this contention: it is true many brilliant kids frequently had trouble in school—at least partially because they didn't colour inside the lines. Many eminent people had a check mark on their report cards next to "Doesn't work and play well with others."

But by my working definition of creativity as the creation of a significant cultural product, this allegedly universal creativity among children is obviously absurd. First of all, children do not create anything significant in art or science. Secondly, it is downright silly to assume that all kids have the potential to be another Matisse, if only adults would encourage green faces and just leave them to their own devices.

What all kids do have, however, and what school certainly does interfere with, is the powerful drive to play. Play, for children, is serious business. Play, for artists, is serious business. This is no coincidence. And every child knows the simple fact that school is work, not play; and work interferes with play. You only put on your play clothes when you get home from school. All the well-intentioned efforts to haul the playground into the classroom are misguided and foolish. Playgrounds already exist. Classrooms serve a different purpose: they are workshops for learning, not creativity. They are places where teachers should be attempting to pass on to their students as much as possible of the accumulated wisdom and knowledge upon which our civilization is founded—and teach these students the necessary skills to use this information. This means (unfortunate as it may be) formal structure and some suppression of individualism. The kid with potential for real creative accomplishment is almost certain to find this a 'hostile environment', but there is no evidence that educational reforms that pay lip-service to 'encouraging creativity' really do so, or create a more 'friendly' environment. If Einstein, mediocre student that apparently he was, had been educated in an "open-concept" school, I doubt he would have been any happier—and he may even have missed learning some critical material that made his discovery of the Relativity Theories possible. But this is not the place for a critique of current educationalist fads. Rather than shooting fish in a barrel, let's return to the idea of art as play.

According to Freud (who took everything very seriously, including play) people always resist relinquishing their pleasures, and if forced to, they transform them, hide them by dressing them up in different guises. Play is serious, and one of the reasons it is serious is that it is one of the primary pleasures of childhood. So what does it get transformed into according to classical psychoanalytic theory? In 'normal' adults, Freud suggests play becomes internalized where it is safe: it becomes the fantasies of daydreams. In artists, it becomes their art. As one writer, Monica Dickens, admitted: "Writing is a cop-out. An excuse to live perpetually in fantasy land, where you can create, direct and watch the products of your own head. Very selfish."*

Assuming this to be true, and it is one of the more plausible of psychoanalytic ideas, how does one reconcile this with the statements of those many artists who insist that the act of creation is hell. One possible, albeit cynical, explanation is that artists, knowing they are among the fortunate ones who spend their life playing, complain about how hard a life it is just to deflect the criticism of the average working stiff.

Another explanation is that the average artist's life really is a hard one. Except for the few that receive wide-spread recognition—and the financial rewards that fame brings—most artists for most—often all—of their lives receive neither the respect the average 'gainfully-employed' receives nor even a living wage. This is the price they pay for playing, and it is understandable that not too many hearts bleed for them. Often, too, the artist's life is hard because of the personal characteristics that come with being creative—the disagreeableness, the hyper-sensitivity, the compulsiveness and socially disruptive monomaniacal commitment to one's art, even (as shall be considered in a later chapter) the serious mental instability.

---

* Monica Dickens is a writer and the great-granddaughter of Charles Dickens. Her autobiography is entitled *An Open Book*.

But probably the major reason many artists are not innocently and childishly happy with spending their lives playing is that somewhere along the way the play has grown to be deadly serious. Certainly children can be serious about their games, but as the years roll by we all become more and more and more excessively serious about everything we do. Adults, even artists, are a bloody serious lot. We set ourselves higher and higher standards and become more self-judgemental—a characteristic uncommon in prepubescent youth. The creative are perfectionists, and perfection is impossible. As William Faulkner said: "The work never matches the dream of perfection the artist has to start with."

IF ART IS A MAN'S NAME, WHO ARE HIS PARENTS?

It is the paradoxical nature of 'serious' grownup play that it really *is* hard work. Why, it is certainly reasonable to ask, do people work so hard for mastery at their play? It is a question that might be easier to investigate objectively, if one looks at what is so unabashedly nothing but play. I refer to sport.

I could be justly accused of bias because of my atypical total lack of interest in sports, but even the sports fanatic has to admit that any objective, disinterested look at sport raises questions about the sanity of the human race. The phenomenal interest in these physical games of all sorts, the incredible personal sacrifices athletes make to master what are after all just made-up games, the fiery passions for these games of both spectators and performers, all would be totally inexplicable to the proverbial alien trying to understand human culture.

Games are invented out of whole cloth, often on a whim. James Naismith, a Canadian Phys Ed teacher at McGill University made up the game of basketball after watching his students idly bounce soccer balls. He took the traditional, tried and true (and tired) idea of putting a ball in a hole, found a peach basket and made up the prototype of modern day basketball. It was a decade later that the peach basket was replaced with holey nets so the ball didn't have to be retrieved after each score. Virtually every existing sport was invented in just such an offhand way.

So we make up games? That's creative isn't it? Some crazed Scot thought it would be fun to see people try to hit a tiny ball with a stick over rough terrain until eventually they knocked it into a small hole in the ground. Another unhinged Scot thought up the idea of throwing rocks down an icy pond at a target circle and so the sport of curling was invented. Well, this is all well and good, for both making up games and playing them is—fun.

What isn't so readily explicable is the seriousness with which these games eventually become imbued. Soccer fans beat each other up with great regularity. A few people really good at a game that happens to be popular are paid millions to play. But most relevant to the current discussion is the dedication and effort that those *not* paid to play show—the long hours of intense effort amateur athletes devote to improving their performance at what is just a silly game someone made up.

·

My daughter, whose primary 'game' is mathematics, once also spent precious hours every day training for collegiate cycling competitions. And the devotion of all those athletes training for the Olympics is legendary, although for only a handful is there ever going to be any remuneration for their efforts. What motivates athletes, from the kid in Little League on up, to devote so much time to what is so fundamentally and deeply trivial and unimportant?

·

Fame and fortune? Not likely. Although every Junior Hockey kid has fantasies about skating for some NHL team before thousands of cheering fans—and taking home six digit paycheques—even Sixth Graders have a better grip on reality than to believe that if they just keep getting up at 6 a.m. for ice time at the local rink, their fantasies will materialize.

·

One doesn't need psychological studies to answer the question. Any guy on the stool next to you at the sports bar can tell you: human beings are naturally competitive and naturally strive for mastery. To an amazing extent it doesn't matter what they are competing about or trying to master. It does not even matter greatly if the competition is with others or with oneself. And it obviously does not matter if the goal one strives for is as meaningless in the grand scheme of things as an athletic victory.

·

All that matters is the evolutionary drive to be good at what one puts one's mind (and body) to. All that matters is our natural instinct to work hard and play hard. And when we play hard, we work hard. This applies to art and science as much as to sport.

THE IMPORTANCE OF ART AND ITS USELESSNESS

As Pop Artist Andy Warhol allegedly replied to the question about what art *is*: "Art? Art is a man's name." If one dismisses sport as fundamentally unimportant in terms of human accomplishment, is not art just as useless? Art is taken very seriously by many people, just as sport is. I take it very seriously. But I have also argued that it is play, and that the serious work associated with this play is somehow parallel to that one sees with sports.

I think that in many ways the artist's motivation is similar to that of the athlete. Both take their play very seriously. Both strive for mastery of what they do just because that is what humans at play do. The artist's competition may more often be with themselves than with others—although, God knows, artists can be notoriously envious of others' successes.

Although the precise function or 'usefulness' of art to both audience and artist is a huge and hotly debated topic far beyond the scope of this book, it is necessary to consider it if one wants to understand why artists take their play so seriously. Just as the philistine considers art at most mere entertainment, the artist considers it the most important and meaningful of all human endeavours. And unlike science—with its usefulness so patently evident in the technology that makes our lives longer and healthier and easier—art's importance is not obvious.

Consider the following frequently proposed 'functions' of art. Many of these overlap.

- Magically controls the environment
- Satisfies a need for exploration
- Satisfies a need for order
- Satisfies a need for stimulation
- Satisfies a need for catharsis
- Satisfies a need for understanding
- Produces pleasing aesthetic experience
- Serves god
- Serves the state
- Communicates

- Educates
- Enriches one intellectually and emotionally
- Entertains

.

Now if you were to ask an artist what purpose art serves, and by implication what purpose he serves, you would receive answers far more poetic than this list. "It is art that makes life, makes interest, makes importance and I know of no substitute whatever for the force and beauty of its process."(Henry James) "Life without art is brutality." (John Ruskin) "Science saves lives. Art redeems lives." (Hippokrites) And I have already argued that art is like science in that both are ways of understanding the world, both are ways of knowing, and together they define human accomplishment and civilization.

.

Replace the word 'art' with 'sport' in this saying: "Life without *sport* is brutality!" The resulting absurdity underlines the difference in the real, lasting value of these two human activities.* However, the objective value of an activity is only rarely related to the motive for engaging in the activity. To drag in another Freudian concept, *rationalization* is probably what most artists are doing when they mouth such noble justifications for their play as "the artist's vocation is to send light into the human heart." (George Sand) A rationalization is not a real reason: it is the excuse we give for our behaviour. It is a pseudo-reason for our actions with which we partially fool ourselves—and try to fool others.

.

The hopeful Olympian who spends at least eight gruelling, usually boring, hours every day in training may say he is doing it for the glory of his country. He may even believe it. But it's a rationalization. His motives are not so noble. So, too, the artist who gives up financial security, sacrifices family and friends, to pursue his art is not really doing it for the glory of his country, his god, or humanity. He is doing it for his own selfish, egotistical reasons. He is skipping school to run off and play.

.

---

* I am not denigrating sport's social usefulness, and I am well aware that an argument can be made for considering it an art. It is true that the rules of, for example, baseball aren't any more arbitrary than the rules of, for example, sonata allegro form in Classical Period music. And the elite athlete certainly shares with the eminently creative many characteristics: talent, developed skills, incredible motivation and devotion. Furthermore, I can see the 'poetry in motion' in many sporting events—at least those I know enough about to appreciate. I just do not see how it furthers our apprehension of the world, which is what art and science do.

One refreshing thing about some creative individuals is that, although they may glorify the value of what they do, they are often quite willing to admit to the private pleasures and selfish motives that drive them. Here are four reasons the writer George Orwell admitted to:

Sheer egoism. Desire to seem clever, to be talked about, to be remembered after death, to get your own back on the grown-ups who snubbed you in childhood, etc., etc. It is humbug to pretend this is not a motive, and a strong one... The great mass of human beings are not acutely selfish. After the age of about thirty they almost abandon the sense of being individuals at all -- and live chiefly for others, or are simply smothered under drudgery. But there is also the minority of gifted, willful people who are determined to live their own lives to the end, and writers belong in this class...

Aesthetic enthusiasm. Perception of beauty in the external world, or, on the other hand, in words and their right arrangement. Pleasure in the impact of one sound on another, in the firmness of good prose or the rhythm of a good story. Desire to share an experience which one feels is valuable and ought not to be missed...

Historical impulse. Desire to see things as they are, to find out true facts and store them up for the use of posterity.

*Political purpose -- using the word "political" in the widest possible sense. Desire to push the world in a certain direction, to alter other peoples' idea of the kind of society that they should strive after.**

Only the third of these four could be considered free of egocentricism. The fourth may sound altruistic, but it is of course predicated on the conviction that one's own way is the *best* way, and one should be able to get the rest of humanity to embrace one's own idea of a better world.

---

* George Orwell's *Essays*. "Why I Write" (1947)

Here are some other totally selfish reasons artists give for persevering in their work:

- Being creative is an inherently satisfying experience
- Pleasure in the exploratory nature of creative work
- Pleasure in god-like power to reinvent the world
- Pleasure in creating what did not exist before but one wishes did
- Pleasure in the exercise of skills
- Pleasure in the final product

And then there is also, especially for writers, the self-indulgent pleasure of letting loose the demon of disagreeableness (that seems characteristic of the creative person) on what one disdains. As novelist Kingsley Amis remarked: "If you can't annoy somebody, there is little point in writing!"

Many things have changed in the last one hundred years, many values have been devalued, and many rationalizations have been discarded. It is hard to pinpoint the moment when art became its own justification and artists stopped claiming that they created for God or country—or to enlighten or to educate. But somewhere along the trail of modern history the tautology "Art for Art's sake" became accepted as reasonable. Many artists no longer felt any need to rationalize their passion. In fact they feel sufficiently confident to even mock their own labours. "All art is quite useless." Oscar Wilde proclaimed—without any apparent diminution of his pride in his own literary accomplishments or self-esteem.

However, the public, the audience, has not been quick to accept this new aesthetic. Along that same trail of modern history that artists trod, the ethically questionable dictum that "If it works, it is good." also became acceptable. We live in an age of unquestioning utilitarianism. So, just as artists gleefully proclaim that art is worthless without feeling any loss of self-worth, the average citizen is more inclined to value the hydrogen bomb because it 'works' than any 'useless' work of art.

Günter Grass expressed this perfectly: "Art is so wonderfully irrational, exuberantly pointless, but necessary all the same. Pointless and yet necessary, that's hard for a puritan to understand." Artists may be serious about their work, but most seem to realize it is play and also realize the value of play to them—and to others. I cannot resist adding one final quotation to this quotation-overburdened section, this from the philosopher Jose Ortega Y Gasset: "Were art to redeem man, it could do so only by saving him from the seriousness of life and restoring him to an unexpected boyishness."

SCIENCE: THE PLAYFUL PLEASURES OF SEEING WHAT HAPPENS

You look out your front window. Two young boys are standing on your lawn by your flowerbeds. One of them has a water glass and is pouring something onto the blooms of your dahlias. The other is watching intensely. You rush out your front door and demand to know what the hell they think they're doing. "Putting salt on the flowers," the one boy replies, wide-eyed. "Bees like sweet things, so we thought we'd see if they like salty things too," the other adds. "Don't be mad, mister," says the first one, "We're just experimenting."

.

Kids try stuff and they call it "experimenting". And of course it is also 'playing' around just to see what happens. And of course this is the basis of all science—and to a great extent all art.

.

However, professional scientists, especially insecure social scientists, vehemently reject this use of the word. It seems to most of them the scientific method is a rigidly defined, rigorous set of statutes, its laws and rules of conduct more immutable and unquestionable than any Church dogma of the Middle Ages. And the enforcement of these regulations is downright Draconian. Because it is a public activity, one's behaviour as a scientist is under constant scrutiny by the most eager-to-convict enforcement agency on the planet—one's research colleagues hoeing the same row.* To breach the rules of research design is tantamount to committing a felony, and there are probably scientists out there who like to see it made a capital offence. He didn't use a placebo in his control group! Off with his head!

.

I teach basic scientific method, and I have a deep respect for it and the importance of rigorous—even what could be called 'rigid'— critical thinking about research. But after berating my students about the importance of rigour in research design, I try to balance this with anecdotes from the historical literature that show that there are two stages to scientific research. The first is very much like the kids with a saltshaker "experimenting" on my dahlias. The second is where the rigour is essential. The average professional scientist, for whom only the latter stage is what they've been trained to consider important,

---

* It is called 'peer review'; the comparison to criminal prosecution by a jury of "one's peers" is obvious.

officially approved dial twiddling of parameters is his primary occupation and preoccupation. While on the other hand, the exception, the highly creative scientist, is too busy playing around (Stage 1) to pay much attention to the rules until it becomes absolutely necessary.

There is a term that historians of science use to describe the way many of the most important discoveries are made: *serendipity*. It refers to the happy combination of chance and a scientist's preparedness. The chance element in this equation is Opportunity knocking—on the lab door. The drudge scientist responds by having the lab sound-proofed so he can continue his dial twiddling undisturbed. The exceptionally creative scientist, out of the curiosity that defines him as a real scientist, walks over and opens the door. Voila!

Examples are legion, if sometimes apocryphal. (Newton may not have actually been literally beaned on the head by an apple.) I'll use two examples, described with only a little poetic license, to make my point. Both are important medical discoveries that also have tangential implications that are somewhat disconcerting.

The first of these is quite famous. In 1922 Alexander Fleming 'accidentally' discovered that lysozyme, a chemical found in tears and mucus, could kill bacteria. Fleming had a severe cold, and while puttering in his lab, his nose leaked, the snot dripping into a Petri dish where he was growing bacteria. The bacteria were almost immediately destroyed! This suggested to him that there might be other substances that could kill bacteria, that there might be a 'magic bullet' for bacterial infection that would not harm the host.

Six years later, Fleming was cleaning Petri dishes that were rife with bacteria by washing them in Lysol, when he noticed that one had a mould on it, a mould that apparently had killed the *Staphylococcus* culture he'd been growing. That the culture had been so contaminated was again an accident, a chance event, which his preparedness and curiosity responded to in an important way. Why were there no bacteria still alive in this mouldy dish? And thus, the first wonder drug, penicillin, was discovered.

Penicillin is a by-product of the penicillium mould that grows well on citrus fruits. The story goes that Fleming had left a mouldy orange lying about his lab and the spores of this are what contaminated his

Staphylococcus culture.* So we have in this moral tale a double example of serendipity. Both discovery of the natural antibacterial agent, lysozyme, and the first antibiotic, penicillin, were not the result of careful trial and error testing of various agents—but rather the result of a runny nose and keeping a dirty lab. And, of course, also of Fleming's attentiveness and preparedness. Neither discovery could have occurred in a modern, hygienic lab, where certainly no mouldy oranges would be lying about, nor where the Petri dishes would be sterilized mechanically, certainly not by the chief researcher.

.

My second favoured example of serendipity is Oscar Minkowski's less well-known discovery. He was using the age-old and crude technique of figuring out what something does by removing it. What goes wrong if you remove the starter from your car? What goes wrong if you remove the front left tire? The wiper blades? The alternator? As these examples show, if you pause to think about them, is that you learn something about what is essential for your car to function, and in some cases may even learn specific functions. But the technique has been quite reasonably criticized for its imprecision. And it is one thing to do this research on your car, where you can put the parts back; it is quite another to do it on animals, where replacement is impossible. So this example is sure to disturb animal rightists.

.

In 1879 Minkowski was removing the pancreas from dogs to see what got broken. One of the unlucky dogs to have had his pancreas removed pissed on the floor of the lab. When Minkowski came into the lab and saw the pool of urine, which was swarming with flies, his reaction was not what would be most people's reaction. He didn't curse out his lab assistant for not having cleaned up the mess, not to mention leaving the window open so flies could get into the lab. He said to himself: "Hmm, this is interesting!"

.

So in this case Opportunity had not bothered to knock, it had come in through the open window, and Minkowski didn't hurry to close the window and clean up the pool of pee on the floor. Instead, his interest was piqued. Why are flies on a pool of urine interesting? Because, as Minkowski knew, urine is not only sterile, it contains no nutrients. Flies swarming about your dog's droppings are not surprising: they are attracted to shit because there is still a meal to be

---

* This may be poetic license; the spores may have come from a mycologist's room the floor below.

had on the not fully digested excrement. Urine, however, normally offers no such dining out. Surely the flies weren't just bathing?!*

So Minkowski analyzed the urine and found sugar in it. Thus, the cause of diabetes was discovered. He lived to see, in 1922, as a result of his important discovery, the introduction of therapeutic insulin, which has saved countless lives.

.

Once again one has to wonder how this fits in with the traditional view of doing science: the methodical, carefully controlled manipulation of variables—with great care to keep extraneous variables out of the picture. What kind of lab was Minkowski running anyway? Dogs pissing on the floor? Open windows with flies buzzing about? (Mouldy oranges in a lab are nothing by comparison.) And one doesn't even want to go near the ethical issues regarding his research that now would cause apoplectic fits in animal rightists. Would ripping the pancreas out of dogs just to see what happens pass a contemporary Ethics Approval Committee for Animal Research? Not likely.

.

Although in science, unlike art, there *are* objective criteria for finally evaluating the 'truth' of a discovery or creation, the actual *doing*, the actual discovering or creating, is much like it is in art. So often, at least in the most important science, the scientist is just like the artist—merely playing around to see what happens. The salted dahlias die. Later carefully controlled studies will conclusively demonstrate the adverse effects of salinity in the soil.

---

* Again, there is another version of the story—one less poetic and less perverse. This version has the dog peeing in his cage, and this behaviour, in a house-broken dog, being what piqued Minkowski's interest in analyzing the urine. But this version seems to me less likely.

THE COMFORTING COMPENSATIONS OF COLD OBJECTIVITY

Scientific creation offers peculiar pleasures unavailable to the artist—peculiar because they are not ones normally associated with creativity. I refer to the satisfactions associated with order and rigor and freedom from emotional elements. There is an authoritarian, even slightly fascistic or militaristic, streak in all science. The best science has no place for petty, popular, emotional human reactions to its creations. (How very different from art, where other folk's emotional reactions to one's creations are so very important!) In science, hey you don't like what I've discovered to be *true*? It makes you feel bad? Tough shit!

.

The examples of this indifference, this deep *disinterest*, of the scientist to the general audience's response to his creative efforts are legion and notorious. Galileo certainly wasn't a big hit with his audience. (Artists who worry—understandably—about censorship should back off and put their concerns in perspective. Arguably scientists have had it even worse.) This isn't to say scientists don't have large egos or that they are totally impervious to the opinions of their colleagues. Where they differ from artists is in their confidence that what they have created has to be eventually vindicated, for it is objective fact. (Such vindication, if it exists at all in art, is only in posterity.) As Galileo knew, the majority of people would have to come around and see the light, have to admit that the earth wasn't the centre of the universe. And those few who didn't could be discounted as fools.

.

When Darwin disembarked from The Beagle, notebook in hand, and said some things that really rattled the *Homo sapiens'* cage, he didn't seem particularly concerned about what his greengrocer thought of his theory. And the opinions of most of his fellow scientists, contrary to the popular misconception, were largely, if cautiously and moderately critical, receptive to his ideas. Scientists *do* attend to criticism from those they feel know enough to offer criticism.

.

The earth isn't the centre of the universe. Man is but one branch on the tree of life that has its roots deep in the primeval ooze. Expression of these insights of scientists about humankind's importance is not flattering and unlikely to endear them to us. Scientists, even more than artists probing the dark side of human nature, seem to mock our species' excessive self-importance and self-esteem. While most folk will now accept on authority the heliocentric view of the solar system, more than half of the putatively literate citizens of America (if polls about such matters are to be believed) haven't yet agreed that other primates belong at the family picnic.*

Just as we human beings once were emotionally committed to a belief in our uniqueness and centrality in the universe and the world of living things, so are we now emotionally, irrationally, convinced of our ability to remedy all problems, convinced of our infinite malleability, a flexibility permitting of perfectibility. Any research that suggests we, just like every other creature on the planet, have instinctual behaviours (such as aggression) is the contemporary equivalent of the Copernican heresy. Almost any research that suggests that genetics, the sexual dice thrown in the conjugal crap game, has more to do with our destiny than our nurturing, our experience, our diet, or our self-improvement exercises is not destined to be a big hit with either the naïve but well-intentioned social reformers *or* the pull-yourself-up-by-your-bootstrap gang. In fact researchers in this area are often subjected to abuse that is downright scary.† No compassionate advisor to a PhD candidate is going to suggest that looking into genetic differences in creativity or intelligence is a good career path to tenure and a peaceful life in Akademe.

Most experienced scientists are quite condescendingly dismissive of social policy crusaders in their field. They understandably feel no need to suffer fools gladly. I remember sitting in our university cafeteria and overhearing a conversation between an eco-activist and a botanist specializing in the Northern forest ecology. The 'tree-hugger' was delivering a sermon on the value of old growth forest and the horrors of evil, corporate greed—epitomized by clear-cutting. Being a bit of a tree hugger myself, I was initially impressed by the passion of the activist. But then the scientist calmly explained some hard facts about the need for old forests, like old people, *to die and fade away* and leave room for young sprouts. He explained how some, albeit certainly not all, clear-cutting methods are actually good for maintaining sustainable forests. He even ventured the expert

---

* They also seem to believe these picnics have only been going on for a mere 6,000 years.
† It isn't only the KKK that burns crosses on heretics' lawns. The vicious harassment by left-wing true-believers of researchers looking at racial or ethnic differences in intelligence or behaviour is embarrassing to everyone who believes in both social justice and objective science.

opinion that we fight too many forest fires, for forest fires too are part of the natural cycle and necessary to maintain an ecologically balanced boreal environment. By stopping the small forest fires that we can deal with, and which serve to replenish the soil, we set the stage for large conflagrations that really destroy the landscape. He made sense and he spoke with an authority obviously based on a profound knowledge of his field of study. But this only made the young activist more impassioned—and eventually downright rude. The scientist eventually stopped responding and listened patiently to the activist's tirade without uttering a word, and when the young man eventually ran out of steam, the scientist smiled, took a deep breath, and then said: "You know, young man, you don't know your aspen from a pole in the ground."

.

Scientists, like many medical specialists, often have a lousy bedside manner. And also like doctors, scientists are often accused of arrogance. Objectively, of course, the real arrogance is that of the ignorant in feeling their own opinions should be treated as seriously as those of the knowledgeable—an unfortunate spin-off of democracy, where the "one person, one vote" principle implies everyone's opinion should be given equal weight. Whatever questionable virtues this principle might have in electing a government, they surely do not apply to science. If they did, there would be departments of astrology and palmistry in every university, and if your two maiden aunts assured you that that funny mole on your nose was nothing to worry about, despite you doctor's concern, you'd not bother with the biopsy.

.

Also, and related to this alleged arrogance, is a particular brand of conservatism. Conservatism is not something normally associated with creativity. In the arts any form of conservatism is thought to be relatively unusual. Openness to experience and conservatism do not seem to go together, and the former is not only a personality trait associated with creativity, it is arguably prerequisite to it. Nevertheless, one form of conservatism is built into training in the scientific method, and so for the scientist it sometimes spills over into other domains.

.

The scientist, like the proverbial Missourian, takes as his motto: "Show me!" As already—perhaps overly—emphasised, scepticism is the cornerstone of all science. In a just court of law, one is presumed innocent until proven guilty beyond the shadow of any reasonable doubt, and it is said to be better to let ten guilty men go free than convict one innocent one. This principle also applies to science, only even more so. Before any scientific relationship is accepted as true, the evidence has to be very, very strong. In fact the rule of thumb in science is that better *twenty* 'guilty' causes go free than one innocent 'cause' be falsely convicted. Scientific journals generally do not even consider publishing findings that are not at least statistically significant at the .05 level. This translates into: "We won't accept that X is causing changes in Y unless statistical analysis shows that there is a less than a one in twenty chance of X being *not* guilty of those changes in Y." As mentioned earlier, this principle is expressed in terms of Type I and Type II errors* or in terms of 'false positives' and 'false negatives', which translate, respectively, into wrongful convictions and wrongful acquittals.

So admittance into—imprisonment in?—the Institution of Established Scientific Facts is difficult. And the corollary is that release from it is not easy either. It is true that equally important a cornerstone of science as scepticism is the acknowledgement of the ultimately tentative nature of all scientific 'facts', of all convictions—and this includes a willingness to release the falsely accused. However, once admission has occurred, release is just as difficult, perhaps even more so in the real world of the scientific establishment. Later this topic and Kuhn's infamous explanation of scientific change in terms of revolutionary "paradigm shifts" will be examined more closely, but what is relevant here is the implications of the inherent conservatism of science for the creative scientist.

It is reasonable to assume that the creative scientist, more than the creative artist, is less concerned with—and perhaps even more uncomfortable with—the emotional side of human nature. Curious as he may be, the status quo—equilibrium, homeostasis, especially in terms of emotion—holds definite appeal. The scientist has been trained in the scientific method and thus inculcated with its inherent conservatism. This may mean that it is much more difficult for the scientist than the artist to innovate, to push the envelope, to challenge the status quo. Artists may strive to upset the apple cart; may even—at least for the last hundred years or so—consider it part of their mission. Yes, the great scientists also, by definition, upset the apple cart, but they only do it incidentally and accidentally (even reluctantly) because their curiosity as to what lay beneath the rickety old cart made it seem necessary.

Furthermore the cold objectivity of science affords a quite comfortable, albeit chilly, shelter—a shelter unavailable in the arts. If what matters most to the creative individual are the results of his creative efforts—and this is almost certainly the case—these results are far more shielded from attack in science than in art. Scientists certainly can and do get tetchy if their findings are challenged, but there is a cold comfort to be found in scientific creativity, a comfort denied the artist. The artist cannot refute his critics, for he has no objective standards he can bring forth to argue for the validity of his productions. He cannot point to empirical evidence and cry, "There, go look for yourself!" At best he can, rather pathetically, point to popularity or notoriety or the admiration of his peers—or (yikes!) the critics. So in the end the artist falls back on his own inner, intuitive conviction that what he has wrought is good—despite what anyone says!

THE PARADOX OF HEART-RENDING ABSTRACTION

Our reaction to abstraction is paradoxical. The continuing aversion of the general public to abstract visual art and that of the average student to mathematics are two pieces of evidence that the human mind does not naturally respond to abstraction. We may value generalities, but we find it easier to grasp them if they are induced from examples, from specificities. It is repeatedly said that only through the particular do we glimpse the universal. The aspiring writer is constantly reminded to "make it concrete", just as the teacher is constantly implored to give examples. Like the devil, the deep truth is in the details.

But having acknowledged this general aversion to abstraction, one has to wonder at the twin paradoxes of music and mathematics. Both are abstract in the extreme. Both, at least in their purest form, are *absolute*—in the sense they are virtually self-contained, independent, and totally self-referential systems with no direct reference to human existence or even the so-called real world.

Yet both seem capable of eliciting the most, profound, deepest aesthetic response of all the arts and sciences—at least for the initiated.

Consider first—for its audience is the largest—music. Of all the arts, it is the one that seems most capable, most of the time, with most people, of producing the strongest emotional, visceral, response. Yet music is the most abstract of all the arts.

Many claim that music is a universal language. Music, it is alleged, speaks to everyone. Yet it has no content: It is as abstract as anything can possibly be. So if it is a language, then what does a Bach fugue *say*? What is a Mozart sonata *about*? They don't tell you anything. They are about pure relationships untainted by any external referent. They are called, most aptly, *absolute* music.*

It is an unresolved question exactly how much training, intentional or incidental, is necessary for a person to respond profoundly to such

---

* Obviously, I'm not talking here about music with lyrics (opera or contemporary rock music) or even program music such at *Pictures At An Exhibition*. This is a topic for later consideration.

pure abstraction, but there are studies that suggest Mozart sonatas produce emotional changes in both animals and infants. This is a fascinating area for empirical studies in aesthetics. But what is germane to the current topic is the indisputable evidence that people, many people, do swear that the deepest aesthetic responses they have ever had are to music, abstract music that has no discernable connection to their human experiences.

So clearly we *are* able to respond viscerally to pure abstraction of certain kinds.

What about mathematics?

For those that confuse mathematics with arithmetic or associate it with bewilderment over those Greek letters that kept cropping up in high school algebra class, my mathematician daughter's claim that some of her most profound aesthetic experiences have come from mathematical theorems must seem incomprehensible. Math, of course, is not arithmetic, nor is it simple problem solving according to memorized algorithms applied to cryptic codes. But it *is* abstract, and it differs from music in that there is no doubt that some training is required to appreciate its beauty. Unlike a Mozart melody, no dog or baby is calmed by being exposed to the Pythagorean Theorem, beautiful as it may be to a mathematician.

Math's beauty, however, is described by the same adjectives as those applied to music—most notably harmony and elegance. Elegance, an aesthetic criterion if ever there was one, is of great concern to mathematicians, as it is to scientists. Mathematicians and scientists frequently speak of the beauty of a theorem or theory. No scientist is surprised to hear a colleague say of a theory "that is too ugly to be true!" Aesthetic value is an unacknowledged criterion for judging scientific (and mathematical) theories. In other words, it is one of the secret collaborators in scientific research.

Probably it's no coincidence that mathematicians and theoretical scientists are noted for their interest in, and appreciation of, music. Einstein with his violin was more typical than exceptional of theoretical physicists. Anyone who has studied musical theory knows how abstract and mathematical it is. Two of the most fundamental elements of music, rhythm and pitch, are most clearly described using basic mathematical concepts. The ancient Greeks used mathematics to discover the basic principles of musical consonance and dissonance, principles that are still the cornerstone of musical composition.

So clearly we *are* able to respond viscerally to pure abstraction of certain kinds. It may very well be that part of the human brain is literally hardwired to respond to the beauty in certain pure abstractions that have specific aesthetic qualities.[*]

So to return to the central theme of this book, one has to wonder why most people and many artists (specifically those who haven't yet ventured across the Great Divide of the Two Cultures) find it incomprehensible that anyone can claim to find a scientific theory or mathematical theorem a beautiful thing, a work of art. I propose the primary reasons for this are the classic two: nature and nurture.

Nature. Our individual brains are different. We are born with different aptitudes that entirely are a function of our physiological makeup. Just as we are born with slow-twitch or fast-twitch muscle fibres, and thus more likely to succeed at either marathons or the 100 yard dash, so too are we born with brains more suited for either concrete or abstract thinking, or perhaps more for verbal than for spatial reasoning, or more for empathetic feeling than disinterested observation. One needs no controlled scientific study to see that this is true. It is obvious to any parent of more than one child that while little Johnnie is clearly good at drawing, little Jeannie is good at solving puzzles. Some people are born with brain configurations that respond well to abstractions. Some are not. And those who aren't will always wonder why Jeannie seems to actually *enjoy* math class.

Nurture. The Two Culture Divide and the dilution of the idea of a liberal education (and, of course, many other factors) have made it easy to avoid dealing with abstraction. The pleasure a mathematician gets from a sophisticated and elegant proof is a pleasure that has to be earned by years and years of preparation. One has to pay one's dues. There often is a very steep price of admission to a great performance. Well, maybe I should say "price of appreciation". If a mathematician shows me what he considers a beautiful proof of an interesting conjecture, I'm like a two-year old at a performance of

---

[*] The right frontal lobe is one of the likely suspects. Functional MRIs and other brain imaging studies, as well as studies of mathematicians and musicians who have suffered right hemisphere brain damage, suggest that the grasp of melody and the abstract visualization involved in higher mathematics are primarily mediated by our right cerebral cortex.

Schoenberg's *Chamber Symphony #1*: I just won't 'get it.' However, while it may take more work to come to appreciate the beauty of science and especially mathematics than it does for the appreciation of most art, both do involve some effort. And within both there is a fairly strong and linear relationship between effort expended and appreciation earned. The difference is in the angle of ascent. The learning curve is steeper in science and math—and the trail is longer. But one does not need to climb to the summit: the view is breathtaking even for those of us already out-of-breath from climbing to even the lower elevations.

.

It is sad so few ever leave the plane, the plain.

CASE STUDIES: RICHARD FEYNMAN & JOHN CAGE

The Nobel laureate physicist Richard Feynman and the avant-garde musical composer John Cage were both notoriously bad boys. They both epitomized serious playfulness. Both have been accused of irresponsible "childishness", but 'child-like' is more accurate.

Richard Feynman was born on May 11, 1918 in Brooklyn and died in 1988. It was in 1918 that Max Planck received the Nobel Prize for his discovery of energy quanta—effectively the birthday of Quantum Mechanics. And just two years earlier Albert Einstein had completed his General Theory of Relativity, which was then to be empirically confirmed just one year after Feynman's arrival on this planet.* So it is interesting that Relativity Theory and Quantum Mechanics, the cornerstones of modern physics, both came into being approximately at the same time as the man who probably was the best equipped to understand them—and elaborate on them.

Many in the know maintain that Feynman is in the running for being the most important theoretical physicist since Einstein. And his intellectual range is even greater. He worked on the Manhattan Project and the development of the atomic bomb; he translated Mayan hieroglyphs; he invented the famous "Feynman Diagrams" for understanding subatomic particle interactions; and he solved the mystery of The Challenger disaster. But he was no desiccated intellectual hiding away in some ivory tower or in the dim corridors of Academe. To call him flamboyant is an understatement.

He may have traded ideas on atomic physics with Einstein and Bohr, but he also traded ideas on gambling with Nick The Greek. He applied the scientific method and his critical intellect to everything in his life from interacting with dancers at a strip club to dealing with charlatans trying to pass off magic tricks as psychic powers. He was an amateur magician and loved magic tricks as much as he hated bullshit, pretence, and junk science. His unflinching critical powers balked at much in current art movements and social movements, yet he was a romantic, whose love for his first wife, who was dying when he married her, is the material for a tearjerker movie. He toyed with the arts in typically irreverent fashion, playing the bongos as

---

* By Sir Arthur Eddington's precise observation of the degree light is bent passing near the sun during a solar eclipse.

accompaniment to ballet and painting a naked female toreador. He picked locks, cracked safes, and played pranks at the top-secret Manhattan Project, driving the security people crazy. In short, he was a real 'character'.

And Feynman is fine—for my purposes—as an example of the archetypal creative personality outlined in a previous chapter: introverted, neurotic, disagreeable, conscientious and open to new experiences. Consider...

Introversion. He was so annoyed at the attention and adulation he received after winning the Nobel Prize that afterwards he would agree to speak only on the condition that his name was not given as the speaker and only the *topic* of his speech was listed in announcements and press releases.

Neurotic. He was so high-strung that his friends found his company as exhausting as they found it stimulating.

Disagreeable. He was notorious for verbally demolishing anyone he felt was presenting intellectually suspect theories. It is said that any scientist presenting a paper who spotted Feynman in the audience, would instantly break into a cold sweat. And if ever there was a person who did not suffer fools gladly, it was Mr. Feynman—despite the fact that he usually presented a smiling, friendly demeanour, had a wonderful sense of humour, and was generally adored by his students.

Conscientious. The extent of his accomplishments speaks for this. Despite the range of his interests, his attention to critical detail is astounding. You don't win the Nobel Prize in Physics for sloppy, careless work.

Openness To Experience. As anyone reading of his life will discover, this may have been his defining personality characteristic.

And many have read of his life. In his very funny quasi-autobiography, *Surely You're Joking, Mr. Feynman,* he recounts the various adventures and misadventures of his fascinating life. This book had co-authors, but he is the sole author of a series of lectures and interviews compiled in *The Pleasure of Finding Things Out.* Both books were best sellers, which says something for their style and charm. (Books by physicists aren't usually best sellers. Stephen Hawking's *A Brief History of Time* also being a notable exception.)

Feynman was also the subject of a best selling biography, *Genius*, by science writer Stephen Gleich,

John Cage was born in Los Angeles on September 5, 1912 (six years before Feynman) and died in 1992 (four years after Feynman), so they are contemporaries, although they lived in different creative worlds. Yet both of these worlds were highly abstract. And both, interestingly, were worlds where the elements of chance played a central role. Einstein insisted God doesn't roll dice, Feynman's quantum theories insisted that He did; and Cage literally rolled dice (or threw the I Ching) to compose much of his music. Both men revolutionized their respective fields: Feynman redefined physical reality; Cage redefined music. They offer an interesting contrast because Feynman was notoriously down to earth, and Cage was almost a parody of the flakey artist. While Feynman was cutting through the bullshit, Cage was flinging it around, albeit sometimes in a most enlightening way.

Cage writes of his early education: "Neither of my parents went to college. When I did, I dropped out after two years. Thinking I was going to be a writer, I told Mother and Dad I should travel to Europe and have experiences rather than continue in school. I was shocked at college to see one hundred of my classmates in the library all reading copies of the same book. Instead of doing as they did, I went into the stacks and read the first book written by an author whose name began with Z.* I received the highest grade in the class. That convinced me that the institution was not being run correctly. I left." This should give one a sense of his personality.

Cage had studied piano as a child, but it was only as a young man that he decided to explore music seriously as his life's work. He then studied with two very different mentors: Henry Cowell and Arnold Schoenberg. Cowell is often considered the father of contemporary avant-garde music because of his total rejection of conventional European rules of composition and his use of tone clusters and percussive rhythms.† Schoenberg, of course, is associated with twelve-tone music and serialism—a highly structured academic

---

* It would be serendipitous and appropriate if he picked Zeno, of paradox fame!
† He 'discovered' tone clusters and their appeal by banging his fists and forearms on the piano. Talk about playing.

approach to classical composition that never really succeeded in capturing an audience. Cage 's music from the 30's and 40's shows the influence of both men, but he soon struck out on a very different path.

Just as Feynman challenged conventional approaches to quantum physics with his 'diagrams' which blurred the distinction between particle and interaction, so did Cage challenge conventional approaches to music with his compositions that blurred the distinction between music and noise.

Tonality in music had already been challenged by his mentors, Schoenberg and Cowell, but John Cage went one very large step further. He said it wasn't a matter of atonality being accepted as musical: it was a matter of discarding the whole concept of music as distinct from any other sounds in the environment, from what was being dismissed as *noise*.

The most radical and infamous expression of this idea was his 1952 composition *4'33"*, which consists of the 'performer' sitting down at a toy piano and not doing anything for four minutes and thirty-three seconds, during which time the audience is forced to listen to the ambient sounds of the performance venue. Each 'performance' will be very different, except in duration, depending on the venue. The one I watched (on video) was outdoors and the ambient sounds included the rustling of leaves and a plane passing overhead. Certainly a performance in a concert hall would be very different, with the rustling of programs and occasional nervous throat clearing being the central motifs of the piece.

Now of course it is easy to ridicule this extreme example of conceptual art. You can actually purchase a recording of a performance, and I, like most people, certainly would wonder why any sane person would spend money on such! As art, it, like much conceptual art, does not meet my criteria for good art. It is unabashedly, appallingly, didactic, even more so than propaganda art, and didactic art is almost always bad art by my standards. Furthermore it is boring, and that is another marker for bad art by my standards.

However, it is important because of what it says. It's a simple lesson, but one that is best taught by the experience. And it is, although a simple idea, a very important one—even if one does not entirely agree with it. Besides, it has the redeeming quality of being funny and outrageous and playful, and perhaps one of the earliest examples of what has come to be called 'performance art'. Only a sourpuss would become indignant or incensed by such serious flippancy. Sensible people will just laugh and say, "Okay, wise guy, I get your point!" And Cage, like Feynman, was a wise guy.

Long before the invention of the synthesizer, Cage was exploring electric and electronic sound sources, using oscillators, turntables, and any sound generating gizmo he could find, including random station-tuning on radios. He introduced some interesting innovations such as his 'prepared piano' where a piano's strings are prepared with nuts and bolts and weather stripping, so as to produce new and interesting sounds. He broke important ground—and long standing musical taboos. He certainly deserves credit for this.

And he did, as well, produce some music that is well worth listening to—and listening to more than once. But he in many ways was more of a theoretician than a composer. He wrote a great deal and has two particularly notable (and poetic) books (*Silence: Lectures and Writings* and *A Year From Monday: New Lectures and Writings*) that I'm probably not alone in finding more interesting and more artistically successful than most of his music. In these he preaches (and practices what he preaches) about the role of chance in composition.

What is particularly interesting about the comparison of these two very different creative men, working in such very different fields, both highly *abstract* fields, is the common personality characteristics they share even though these common characteristics are manifested so very differently. Both were playful to the extreme—and both drove their stern and serious and 'mature' colleagues up the wall with their antics. Only their methods and fields of endeavour were different.

Both challenged conventional, ingrained ways of thinking in their respective fields, and certainly both deserve to be called critical thinkers. But Cage's mind was far from logical, and he was quite uncritical in embracing whatever struck his fancy. (He apparently took the *I Ching* seriously—or at least as seriously as he took anything.) Feynman, on the other hand, apparently couldn't put aside his critical training as a scientist and see behind the actual empirical gesture when it came to dealing with art. (His comments on modern poetry are embarrassing.)

So while in one sense Cage and Feynman are virtually archetypal of artist and of scientist, two beings from different worlds, I think they would've got along swimmingly. They'd have instantly recognized the twinkle in each other's eyes.

COMMON GROUND AT THE CONFLUENCE OF ART AND SCIENCE

"By keenly confronting the enigmas that surround us, and by considering and analyzing the observations that I have made, I ended up in the domain of mathematics, Although I am absolutely without training in the exact sciences, I often seem to have more in common with mathematicians than with my fellow artists."
—M. C. Escher

"The ideas in the Large Glass are more important than the actual realization. The "Large Glass" constitutes a rehabilitation of perspective. For me, it's a mathematical, scientific perspective, based on calculations and on dimensions. Everything was becoming conceptual, that is, it depended on things other than the retina. What we were interested in at the time was the fourth dimension. Simply, I thought of the idea of a projection, of an invisible fourth dimension, something you couldn't see with your eyes. "The Bride" in the "Large Glass" was based on this, as if it were the projection of a four-dimensional object. I called "The Bride" a "delay in glass." A tactile sensation which envelops every side of an object approaches a tactile sensation of four dimensions. Consequently the act of love as tactile sublimation could be felt as a physical interpretation of the 4th dimension."
—Marcel Duchamp

Some artists were well aware of science as a 'secret' agent in the creation of art and acknowledged it publicly. But many used it and didn't even think of it as science. And scientists similarly seemed not to realize how much influence art, and especially aesthetic values, had on the way they thought.

WHERE SCIENCE AND ART CONVERGE

The 20th century gave us many major artists who certainly did not view science as the enemy—Escher and Duchamp being excellent examples. And that century's most notable scientists and mathematicians have repeatedly emphasized the importance of aesthetic evaluation in judging their own creative products.

.

So not all artists and scientists lived far upstream along different creative tributaries, and those who didn't may offer some hints to future currents and confluences downstream. The third and last pane in this triptych examines what I believe will be the future convergences of the two creative streams, but first here is a brief, upbeat look at current currents.

.

It is technology, more than science, which has seduced today's artists. Technology was embraced by artists largely because it gave them new tools to create and distribute art, for that of course is what technology is—tools made possible by science. Many may not have been sufficiently sophisticated in science to take inspiration from the new ideas, but they were more than willing to use the tools those ideas made possible. And also, just as technology follows from science, so does some grasp of science follow from learning new technology.

.

Consider John Cage, who (except for his interest in mycological taxonomy*) is about as far from a scientific mind as one can imagine, but was one of the pioneers in the use of technology in musical creation. Jean Cocteau, Salvador Dali, and other surrealists quickly embraced new filmic technology to create their visual art. (When, in 1929, Hollywood was making *The Broadway Melody*, Cocteau was making the bizarre film classic *Un Chien Andalou*.)

.

And writers used the new technology in an even more fundamental way—as the means to distribute their writing. New inexpensive reproductive technology† such as the mimeograph machine made

---

* One of Cage's passions was mushrooms, and he was quite an expert on the taxonomy of mushrooms—or, given his fondness for 'hunting' and eating them, he wouldn't have lived as long as he did.

† I'm not referring to artificial insemination or cloning! The development of technology to reproduce art (be it written, auditory or visual) is of even more importance.

possible the publication of work outside whatever establishment was controlling conventional publication. The term *Samizdat* is a compound Russian word for 'self' and 'publish' coined in the late 1950's when mimeographed writings were secretly distributed in the repressive Soviet regime*, but even in the 20's mimeographed writing was so distributed. In the West, particularly North America, the literary renaissance in poetry and radical literature was made possible by 'small presses' which relied on cheap reproductive technology to bring their works to a relatively small but very engaged group of readers.

Other contemporary examples of the use of technology in the arts are numerous, but more unusual are the use of actual science, specifically the *ideas* of science, in the arts. However that too happened and seems to be happening more and more often.

In music, exploration of new technology led composers into the complexity of scientific acoustics. Many composers who now use synthesizers can comfortably discuss Fournier transformations with physicists and mathematicians. Highly innovative American composers like Babbitt and other serial and 12-tone composers treated musical composition as a science, turning to early psychophysicists and researchers into auditory perception, such as Helmholtz, as their guides to composition. Simultaneously they rejected the 'unscientific' European musical tradition with its strict, formal guidelines to 'correct' musical composition.†

Visual artists who have become enamoured of fractals have been lured into abstract mathematical territory and can actually talk intelligently about limits and iterations and strange attractors. The complex physics of ray-tracing is another area in which artists,

---

* Sometimes the distributed material was in the form of carbon copies of completely retyped works, because even mimeograph machines were carefully guarded by the KGB. Reproductive technology is subversive.

† Babbitt carried this so far that he became infamous for implying (in a controversial article for *High Fidelity*) that the average listener to this (his) new 'scientific' music couldn't expect to understand and appreciate it—and shouldn't even bother to try, for it required a highly trained ear, just as appreciation of the beauty of a complex mathematical proof of theory in modern physics requires extensive training. Naturally, there was a reaction to this misguided relegation of serious music exclusively to Academe. It was called 'Minimalism', and its practitioners (composers such as Reich and Glass), although still drawing from scientific ideas, do not produce esoteric music. In fact, they enjoy wide popularity. Philip Glass is said to be the most prolific and successful composer of operas since Rossini!

seduced by the potential of computers for creating art, have become knowledgeable. I have a friend who, for years, created his visual art by writing computer code to generate ray-traced images.*

The performance arts have been most notably affected more by technology than actual scientific ideas—with the advent and virtual take-over of film as the primary medium, and live theatre largely relegated to a pricey entertainment and social event. But even film-makers have moved beyond mere concern with how technology can be used to create spectacle into the deeper realm of pure science when they venture into the ideas of science, into what is originally a literary domain: science fiction.

Which brings us to the one art form that has been truly transformed by science *as* science: literature. Scientific ideas have spawned one of the most popular literary genres of all time—science fiction. The best science fiction is idea fiction. It is in fact a very highbrow genre, despite its reputation as juvenile literature. It no doubt got this reputation from its popularity with the young, but it is slanderous that it is not more widely recognized as 'serious' literature. The condescending dismissal of science fiction by conventional literary critics and commentators has more to do with awareness of the immaturity of many of its fans than with awareness of the genre itself.

Its appeal to youth is quite explicable. The young are attracted to ideas more than to felicitous use of language—and certainly more than to character and complex motivation and subtle internal moral conflict. The young have too little experience of the world to appreciate such matters. But they *are* captivated by raw ideas and have the intellectual capacity to appreciate them.

A sixteen-year-old is moved by the primitive, hormonally derived emotions of the most appallingly saccharine pop tune. But he is also moved by the purely intellectual ideas that moved and motivated our most mature and sophisticated scientific minds. No one assumes a sixteen-year-old knows more about interpersonal relationships than even a socially inept fifty-year old. But many a lad or lass that age knows more about scientific ideas than those far more experienced in the real world. The old method of calculating IQ was based on

---

* The software for doing so has improved so much that one no longer needs such mastery of the basics—although even now it still helps.

dividing mental age by chronological age—up until the age of 16! At that point one's intellect (but not, of course, one's emotional or moral maturity) was assumed to be equivalent to that of any adult. This is of course ridiculous.

It is true that much science fiction is poorly written even when it does confront important ideas. It is true that much science fiction has character development at the level of the worst TV sitcom. But it is also true that much highly esteemed 'serious' literature has virtually no real intellectual content and is, aside from skill in use of language and effectiveness in creating character, not fundamentally different from an afternoon TV soap opera.* And there are those shining examples of science fiction that meet and beat the criteria applied to both conventional literary fiction and science fiction, works by writers that have been claimed by the conventional literati as their own, and not 'really' science fiction writers—people such as George Orwell and Aldous Huxley, or more recently Kurt Vonnegut, Margaret Atwood, Ray Bradbury, Arthur C. Clarke, Ursula K. LaGuin, and Philip K. Dick.

So science (and its offspring, technology) have to some extent invaded the arts. Is the converse true? Have the arts influenced scientific development?

I would say yes—but to a lesser extent and at a far more abstract level. Scientific theories are metaphors, and insofar as scientists find in the arts appropriate metaphors they appropriate them. Furthermore, it is my impression that scientists, on average, know more about the arts than artists know about the sciences. So when a scientist is searching around for an appropriate metaphor in which to capture an idea, he has the knowledge base, the resources.

In short, the two cultures are probably not as divided as they were when C. P. Snow coined the phrase. Some say the world has *shrunk* and speak of Marshall McLuhan's 'global village'. Some say the world has *expanded* as we become less parochial and ethnocentric and made more aware of the world's incredible diversity. Both are in some sense true. And both mean that inevitably scientists and artists are no longer living in separate isolated villages. But when two cultures are thrown together, conflict is inevitable.

---

* I won't name names.

EDDIES AND TURBULENCE CURRENT AT THE CONFLUENCE

Scientific progress is a mixed blessing, and those that focus exclusively on the negative may even call it a curse. While it is obviously inaccurate to claim that all art resides on high moral ground, many artists seem to believe this—and can be quite self-righteous and sanctimonious about the evils scientific knowledge has made possible. A disconcerting number of artists seem to endorse implicitly the idea that Adam should've left that damn apple alone.

.

Ethics and aesthetics are two different branches of axiology (the philosophical domain that deals with value judgments). However they are not congruent, as evidenced by the Nazi SS aesthetes who loved their opera or the writers like Celine and Pound who supported fascism. Art is no more *inherently* moral than science. Both are ultimately *amoral*. Both can be applied to good or evil ends. The artist's skills can be used to manipulate human emotions and the scientist's to manipulate the physical world. Such powers are inherently dangerous. The atomic bomb is the most obvious example of the terrifying power of science. But it is the propagandist's art that makes possible the public, democratic endorsement of such misapplication of power as the bombing of Hiroshima and Nagasaki. Fortunately it is probably true that the majority of both artists and scientists are highly principled people.

.

So I think it is fair to say that the Two Cultures are no longer so much separated by ignorance of each other as they are in conflict because of being forced to interact in a situation that makes them wary of each other's motives. If Keats was concerned with science destroying the magic of the world, many artists now seem far more concerned with science *literally* destroying the world. And scientists now are not so much ignorant of art as antagonistic to artists who seem to be opposed to the pursuit of knowledge.

.

An interesting irony of this is evident in conceptual art, where much contemporary work uses the technology made possible by science to attack science! Multimedia productions, using the latest audio-visual gizmos, often take as their theme the alleged sterility of our contemporary hi-tech life-style. It wouldn't surprise me to see a sky-writer scribbling on the air a message about conserving fuel.

.

But probably the most interesting and paradoxical and conflict-ridden 20th century confluence of science and art is the young discipline called psychology. But that complex issue is the subject of the final section of this first pane.

CASE STUDIES: ARTHUR C. CLARKE & LOREN EISELEY

Arthur C. Clarke and Loren C. Eiseley are two brilliant and influential writers who derived their themes from science but whose primary concern is the human condition. Clarke is a writer of fiction and Eiseley was primarily an essayist, but they have much in common—most notably a profound sense of wonder at the nature of the universe. They do, however, differ in where they usually stare in wonderment and awe. Clarke, who holds a degree in physics and mathematics, looks skyward and to the future. Eiseley, who holds degrees in geology and anthropology, more often looks earthward and to the past. Yet their vision of man's place in the universe is very similar—and that vision is of a quest. It is not pure coincidence that Arthur C. Clarke's best known work is *2001: A Space Odyssey*, and Loren Eiseley's best known book is *The Immense Journey*. Both have as a central theme the evolution of Man.

.

Arthur Charles Clarke was born in Minehead, Somerset, England on December 16, 1917. In 1936 he moved to London, where he joined the British Interplanetary Society and began to experiment with astronautic material. In 1945 he published his most important scientific contribution, a paper on satellite communication, *and* his first work of science fiction.* Many books followed, both science fiction and scientific non-fiction. In 1956 Clarke took up residence in Sri Lanka (then called Ceylon) where he continued his prodigious output.

.

Some years have special significance, and 1968 certainly qualifies. It is usually considered the epitome of "The Sixties". It was the golden age of hippies and yuppies. The Beatles released their *White Album*. It was the sex, drugs and rock 'n roll. But there was a dark side to the year. Martin Luther King was assassinated and major riots followed in many American cities. Robert Kennedy was assassinated, as was Andy Warhol. Viet Nam was coming apart at the seams, and the anti-war movement was coming to a crescendo. The Chicago police, gone ballistic, sent over a hundred protesters at the Democratic Convention to the city's emergency rooms, arresting nearly two hundred others.

.

---

* "Extra-terrestrial Relays" and "Rescue Party", respectively.

And then, at the end of the year, the first U.S. mission to orbit the moon, Apollo 8, was launched.

That is also the year that the film *2001: A Space Odyssey* was launched. This film and the novel of the same name were the result of collaboration between Clarke and the great film director Stanley Kubrick. But unlike most of Kubrick's other films, the writer stands on at least equal footing with the director.* I remember when it came out and all the debate about what it "*meant*". The ambiguity and subtlety of this work left no doubt in anyone's mind that this was at least intended as high art. This was no space opera. This was not *Invasion of the Body Snatchers* or *The Blob*. It had more in common with the works of Bergman or Antonioni—or James Joyce—than pulp science fiction or Hollywood flicks.

So what are its themes? What is it about? It is too complex a film for me to even attempt any detailed 'interpretation'. But like most important art, it has generated a substantial body of critical commentary based on a close 'reading', and, for those interested, at the time of this writing, there exists an excellent 'review' of the film by Tim Dirks available on the Internet.† Dirks presents a detailed, thoughtful, in-depth, and very readable explication. The review begins with the statement that "*2001: A Space Odyssey* is a landmark, science fiction classic—and probably the best science-fiction film of all time." I'm sure many would disagree with this assessment, for even without close analysis it is obvious that in this film we have true merging of the two cultures. Science and art come together here in a profound way. This isn't science fiction because it is set in the future (or what was then the future!). No, this is fiction that is literary fiction where scientific knowledge and mystery—and not mere technology—are at the heart of its aesthetic impact. The human journey, including the mystery of evolution, are its central themes, but along this journey such troubling questions are raised as to the nature of intelligence and sentience—and the place, the significance, of us piddling humans in the vastness of the universe. And these are the very questions that science, with its unsettling predilection for

---

* *The Shining*, another great Kubrick film, is typical. It is 95% Kubrick and 5% Stephen King, who allegedly liked it least of all the films made that were based on his books—even though film critics consider it the best.

† The url of this review is: http://www.*filmsite.org/twot.html* Unfortunately, given the transitory nature of most websites, the reader may find this site has disappeared into cyberspace ether.

diminishing our self-respect, with its discoveries about evolution and cosmology, has repeatedly made us confront.

.

Clarke had been confronting these questions in his fiction for many years before this film made him famous (e.g., in *Childhood's End*, which was published in 1953), and he continues to grapple with the philosophical and social implications of new scientific knowledge. A few years ago I attended a conference on "Creativity in Art and Science" and Clarke was a guest panellist via—most aptly—satellite connection from his home in Sri Lanka. In 2005 he published *Sunstorm,* in which he continues to explore complex philosophical themes by projecting the reader into the future.

.

Loren Corey Eiseley, like Clarke, was born into a family less than well-off. He grew up in a rural environment, coming to formal education relatively late. But the world Eiseley came into in 1907 was not the domesticated English countryside of Clarke's childhood; it was the great plains of Nebraska. His family lived on the outskirts of town, and his childhood entertainment was wandering the prairie countryside, exploring creeks and ponds and caves—and developing the love for, and knowledge of, the natural world that permeates his writing.

.

Eiseley's mother was deaf, an artist, and emotionally unstable. His father worked as a hardware salesman, but was also an amateur Shakespearean actor with a love for language. It is of course facile psychologizing, but nonetheless tempting, to attribute Eiseley's often melancholic nature to his mother, his love of language to his father, and his retreat into nature to discomfort with his parents' dysfunctional relationship. No matter what the influences, Loren Eiseley was drawn to the mysteries of the natural world and driven to contemplate our place in it. The result is truly magical prose.

.

His essays are masterpieces of the genre, and he is one of the great stylists of the 20th century. They are about being human, but they often take as their starting point the non-human world, the naturalist's world. He unites science and literature in a very different way than Clarke. For science does not lure Eiseley to speculation about the future; it lures him into the past. He was a bone-hunter, and the fossil record was his Bible, his guide to understanding our species' place in the grand scheme of things. "Every time we walk along a beach some ancient urge disturbs us so that we find ourselves shedding shoes and garments or scavenging among seaweed and whitened timbers like the homesick refugees of a long war."

His first, and probably best-known, book, *The Immense Journey*, was published in 1946. This collection of personal essays has as its major thread the immense journey of evolution. He went on to write several books on Darwin and evolution, for the wonder of the human journey clearly obsessed him.

He also wrote a book on Francis Bacon (the man often called the father of the scientific method) in which he demonstrates his deep understanding of what science is really about, for Eiseley was a man of science, but not of the dial-twiddling sort. As he wrote in his essay on "Science and the Sense of the Holy"*: "In the end, science as we know it has two basic types of practitioners. One is the educated man who still has a controlled sense of wonder before the universal mystery, whether it hides in a snail's eye or within the light that impinges on that delicate organ. The second kind of observer is the extreme reductionist who is so busy stripping things apart that the tremendous mystery has been reduced to a trifle, to intangibles not worth troubling one's head about."

In his various collections of familiar essays, he often takes as the starting point to his meditations some other species. He had an intuitive grasp of our relationship to other living things and so was able to follow a fine but very strong thread from any creature, be it bat or pigeon or fern or spider, back into the complex web of life. As he writes in the marvellously titled *The Unexpected Universe*: "One does not meet oneself until one catches the reflection from an eye other than human."

In the early seventies he published two volumes of verse.† His own description of one of these books, *Notes Of An Alchemist*, could very well describe his art in general and why it stands as a beacon to guide those venturing into the turbulent waters at the confluence of art and science: "{My writing is} a kind of alchemy...by which a scientific man has transmuted for his personal pleasure these sharp images into something deeply subjective." And it must be added—something deeply, aesthetically beautiful. It is not surprising that his audience in

---

* *The Star Thrower*
† These works do not display the same mastery of poetry as he had of prose, but do have wonderful moments.

both the scientific and literary communities has continued to grow—
to *evolve*.

.

Loren Eiseley came to the end of his own personal, immense journey
in 1977 at the age of 70. His tombstone bears an epitaph that is a line
taken from one of his poems: "We loved the Earth, but could not
stay."

QUICKSAND ON THE SHORE: THE SOCIAL EVILS OF SOCIAL SCIENCE

"The value the world sets upon motives is often grossly unjust and inaccurate. Consider, for example, two of them: mere insatiable curiosity and the desire to do good. The latter is put high above the former, and yet it is the former that moves one of the most useful men the human race has yet produced: the scientific investigator. What actually urges him on is not some brummagem idea of Service, but a boundless, almost pathological thirst to penetrate the unknown, to uncover the secret, to find out what has not been found out before. His prototype is not the liberator releasing slaves, the good Samaritan lifting up the fallen, but a dog sniffing tremendously at an infinite series of rat-holes."
—H. L. Mencken (*A Mencken Crestomathy*)

.

"*Psychologists are wannabe artists and wannabe scientists. Of course virtually none have attained the former goal and relatively few have attained respectable status as the latter. That many of them also want to be wise men and do-gooders, suggests that they are totally out of touch with reality and that the whole field is infected with a virulent case of hubris one can only hope is terminal.*"
—Hippokrites

Psychology may *seem* to be at the very confluence of science and art. It is largely a 20th Century phenomenon—and its popular success is as phenomenal as it is scary.

.

I started university as a major in biochemistry. My reasons for choosing this area of study were simple: I liked biology and I liked chemistry, so I thought I'd become a biochemist. I also harboured a desire to be a writer, but even then I had the sense to realize that studying English wasn't a sensible path for a wannabe author. I'd already had enough experience with the academic study of literature in high school to know it was more likely to damage than advance this—to my mind at the time—outrageous ambition. Science, on the other hand, was something one really could learn in school.

.

I soon found I had neither the aptitude nor sufficient interest for the study of chemistry or biology. (Maybe, too, I was just distracted by too many other things—everything from mini-skirts to literary fiction. Plus I hated slide-rules.) So again applying the same dubious reasoning that misled me into biochemistry, I figured I'd switch my major to an area I thought combined several of my current interests. I was interested in science and I was interested in philosophy and I was interested in the humanities. So, naïvely I switched my major to psychology. This seemed—to my inexperienced mind—a discipline at the crossroads between science and the humanities.

.

I now tell my Intro Psyc students, on the first day of class, that if they really want to understand human nature—understand why their father is a drunk, their mother a whore, their lover a liar—they should drop psychology and take a course on Shakespeare: the course instructor may or may not be enlightening, but the Bard could teach them much more than any 'professional' psychologist.

.

But when I wandered, staggered, into the study of psychology, I was young and foolish. I didn't know that psychology, rather than combining science and art, was actually a pretender in both areas—not really a science and certainly not a field that could legitimately claim to be part of the humanities. In fact, psychology is largely hokum—which would be okay and tolerable if, like astrology, otherwise intelligent people didn't so often take it seriously. Unfortunately, it is taken seriously, far more seriously than it deserves, and probably has done more damage than any other pseudo- or quasi- science—except perhaps sociology or political 'science'.

.

I will add a small qualification. Psychology, when really practiced as a science, *has* discovered some significant things. Many of these are more in the domain of biology than psychology—but not all of them. We do understand more about learning and memory and even a little bit more about social interaction as the result of psychological research. And, as a spin-off, probability theory (so essential to drawing any kind of conclusions from a science that yields 'sloppy' data) was greatly advanced by that mother of invention—necessity. It may even be that some of the research on creativity and the psychology of art really might have something to contribute. Obviously, I must think so—or I wouldn't be referring to it in this book

.

Anyway, having openly stated my prejudices, it is time to take a closer look at psychology—the discipline that I once believed could bridge the gap between the Two Cultures.

THE ABDICATION OF PHILOSOPHY

Ah psychology, that embarrassing 'science'!* It is difficult to explain how psychology and sociology rose to their current elevated status. It is very complicated. There so many obvious contributing factors: the rise of democracy and decline of the Church and the aristocracy; the demonstrated efficacy of empiricism; the elevation of the individual over the community; the replacement of canon law with civil law; the dramatic increase in narcissism that leisure allows.

It is sometimes said that the three giants of Classical Theatre—Aeschylus, Sophocles, and Euripides—represent a rapid progression from concern with the spiritual, through concern with ethics, to concern with individual psychology. (Euripides is as modern a playwright as Aeschylus is incomprehensibly foreign to contemporary sensibilities.) In some simplistic sense this progression is reiterated on the grand scale of Western Civilization. The Middle Ages, where God was the only real creator and the artist a mere artisan serving Him, was followed by The Enlightenment, where Man could also be creator. As the power of the Church waned and modern science was born, ethics became an independent discipline. Philosophers as diverse in character as Kant, Hobbes, Rousseau, Voltaire and Locke all turned their attention to ethical issues, to the central question of how to live one's life—and the meaning of life. These were the fundamental philosophical questions addressed by Plato and Aristotle, but that was before the rise of The Church as ultimate moral authority. With the Enlightenment once again the answers to these questions became open-ended, no longer delivered nicely packaged and indexed by The Church. And, as The Church realized, these philosophers were walking along a slippery slope, with empirical science watering the incline.

Eventually intellectuals did slide down from the secure heights of religious dogma into what many consider the abyss of Modern Philosophy, ending up either in the quagmire of existentialism or the rocky pit of logical positivism. In one sense, after Sir Francis Bacon and the subsequent triumph of empiricism in epistemology, it was all downhill not just for religion but for 'pure' philosophy as well.

---

* I'm being kinder than William James, the philosopher who wrote the first Psychology text. He called psychology the "nasty" science.

Loren Eiseley says of Bacon, who is often considered the father of modern science, that he "...more fully than any man of his time, entertained the idea of the universe as a problem to be solved, examined, meditated upon, rather than as an eternally fixed stage, upon which man walked." While Bacon's conception of science is a far cry from its contemporary incarnation, the change in worldview that followed was to undermine traditional philosophy just as philosophy had undermined traditional religion.

.

It is a truism that science has invaded and conquered much of the once vast empire of philosophy. Many, maybe most, of the questions that once obsessed philosophers have been usurped by scientists. We may still honour the philosophers for first asking the questions, but we now turn to science for the answers. The nature of matter? Good question, Democritus, glad you asked! We turn to the physicist. The origin of the universe? Good question, Lao Tzu! We turn to the astrophysicist and the cosmologist for the answer. The origin of life? Good question, St. Thomas! We turn to the biologist and chemist for the answer. The meaning of life? Good question, Sartre! We turn and turn again.

.

The meaning of life? We admit that it is not an empirical question. It cannot be answered by scientists nor by deductive reasoning, and so perversely we decide it is a 'meaningless' question as outside of the domain of philosophy as it is outside the domain of science. We chose to live in the rocky pit of logical positivism with A.J. Ayers.

.

The meaning of life? We admit that it is not an empirical question. It cannot be answered by scientists nor by deductive reasoning, and so perversely we decide life is 'meaningless'. We chose to dwell in the quagmire of existentialism with Jean Paul Sartre.

.

Of course this is a grossly unfair and simplistic characterization of the two dominant philosophies of our time, but it is *not* unfair to say that philosophers have largely abdicated their responsibility to deal with this paramount question. And since the question will not be so simply exorcised by being ignored, someone eventually steps forward with some 'answers'. He carries empiricist credentials, even calls himself a scientist. Enter stage left, The Psychologist!

.

By the middle of the nineteenth century, science had so impressively demonstrated its ability to answer difficult questions about the physical world that it was only natural that it should turn its attention to the nature of man and his mind. To a man with a hammer, every problem is a nail. Science is one helluva powerful hammer. If empiricism can explain why apples fall, why shouldn't it be able to explain why minds arose? Thus, sociology and psychology were born. Both were from the start wannabe natural sciences. Even Freud with his amorphous concepts of id, ego, and superego believed in these hypothetical constructs as actual brain structures yet to be discovered. And, as a medical doctor, he certainly considered himself trained in science.*

The most universal of empirical experiences is our own mind, the constant flow of impression and thought—and the intense awareness of self. It was the reason for Descartes believing he was really grounding his philosophy on the most fundamental, indisputable, *empirical* bedrock possible when he declared "I think, therefore I am!" Empiricism is based on observation, so early psychologists tried the most obvious approach to understanding mind: they tried to observe it directly. The movement was called *introspection*.

It didn't work. Try it. Try 'watching' yourself think. It's the most confusing house of mirrors. Science is objective, so, really, how could subjective experience possibly be objectively observed? It would be wonderful if we could observe ourselves coming up with ideas, but we can't—for ideas seem to just appear without leaving any detectable trail. It is no wonder that the Greeks hypothesized The Muses as messengers delivering ideas to the lucky ones—the ones we now call creative, and were called, apropos the Greek view of the process, "*inspired*".

The psychoanalytical school continued to try to understand the mind, including the mind of the creative, using indirect probes such as free association and dream analysis, but mainstream psychology, paralleling the logical positivists' cop-out, decided to censor 'mind' as a dirty word to be banned from public discussion. Radical behaviourism was indisputably scientific—and also, unfortunately, indisputably narrow in its focus. But it isn't very informative.

The whole of human behaviour, everything from licking an ice cream cone on a hot summer day to penning the *Tropic of Cancer* in lonely, heart-broken exile, can indeed be described by this simple paradigm: Stimulus-Organism-Response. S-O-R. Stuff comes in, something

---

* Some would say this particular delusion persists among most doctors today.

happens inside the organism, and stuff comes out. The behaviourists knew that stimuli were physical phenomena, measurable, the nitty-gritty stuff that traditional science can deal with. As were responses. It was that damn black box, the organism, and specifically the mind of the organism, which was impossible to probe and measure. So: hell with it! Let's just excise the O from the S-O-R schema. S-R theory was born. Find predictable relationships between the measurable input, stimuli, and the measurable output, responses. Good science. But bad news for those who think thinking matters.

.

Behaviourists did discover much that was useful, but like logical positivism it held no interest for those who wished to understand what is uniquely human—our minds, our ability to create, to dream, to ask questions and seek answers. One would have thought that people would turn to the arts, specifically literature, for certainly there is more psychological insight of the type we humans care most about in the plays of Shakespeare or the essays of Montaigne than in the graphs and statistical analyses of response rates to different reinforcement schedules imposed on Wistar Strain rats pumping a lever for food pellets in a Skinner Box.* But science had a hold on the public imagination.

.

So those psychologists with humanist interests stepped forward, and because they could claim to wear the mantle (i.e., lab coat) of scientists—for psychology was a science, was it not?—they had the double blessing of being (supposedly) scientific *and* concerned with what most people are concerned with—their inner lives. They came over from clinical psychology and revisionist psychoanalysis. Sometimes they were, unfortunately, often the same people that in a non-secular situation would be Evangelical Christians.

.

And this is when things started to get ugly. Okay, just silly most of the time, but ugly only too often.

---

* S-R theory is good at explaining why slot-machines keep people pumping in quarters, but it really has little to say about why writers keep on writing or readers keep on reading.

HOW SOCIAL SCIENCE HAS MUDDIED THE WATERS

Philosophy originally concerned itself with understanding the universe and with understanding our place in it—the experience of being human. The former of these concerns is now largely the domain of physical science, and the latter is, far less appropriately, now considered by most people to be the domain of psychology.

When I ask my Introductory Psychology students for their definition of psychology, they almost invariably reply: "The study of the mind." To which I reply: "That's a good definition of the *humanities*, of what philosophers and historians and artists do, but frankly it isn't the definition 'professional' psychologists apply to their discipline. The current, almost universally accepted definition is 'the *scientific* study of *behaviour*'." The key words here are 'scientific' and 'behaviour'. (Only lately, with the decline of behaviourism, have psychologists started to once again speak the forbidden 'm' word, so they feel they can dare claim they study behaviour *and* the mind.)

Generally students interested in understanding themselves, their minds and their behaviour, as well as understanding other people, do not sign up for Philosophy 101. Nor do they select literature courses where they will be exposed to the great novelists and playwrights and essayists who could really enlighten them. Instead they register for an Intro Psyc course. And the numbers support this contention.*

They of course are disappointed and probably don't walk out of the final exam with very much more real understanding of their mothers or lovers—or themselves—than if they'd taken Business Writing or Geomorphology. At one end of the spectrum, psychology blurs into biology, specifically neuroscience. At the other end of the spectrum, the social psychology end, it blurs into sociology. And sociology is even more frightening than psychology, for it is even less based on real science and more influenced by political correctness than empiricism.

Perhaps the central problem with understanding the so-called 'field' of psychology is that it isn't really an independent discipline, but rather is like a politician on the campaign trail, trying to be all things

---

* Last year my Intro Psyc course had the largest enrollment in the university, and this is typical of liberal arts colleges.

to all people. And, like the politician, consistently fails to deliver on its promises.

.

It is a constant source of embarrassment to me to confess that I earn my daily bread teaching a subject I think not really a subject at all. Most of what is of real value in this putative discipline is more accurately called biology or neuroscience. Yet what interests most people (including, alas, most of my students) about psychology is material that in a reasonably organized library would be filed under philosophy, anthropology or self-help manuals. As philosophers, most psychologists are pretty sophomoric. As anthropologists, they are even more naïve than most real anthropologists. And as guides to making ourselves better human beings, they are the blind leading the blind—while pretending to have 20-20 vision.

.

I'm over-stating my case a bit, I admit. And since in this book, I do repeatedly refer to so-called 'psychological findings', I'm undermining some of my own arguments by such a blanket dismissal of psychology. However, perhaps by now examining various aspects of psychology, I can temper my remarks *and* still support some of them.

ABUSES, MANY

As I've already suggested, a very good case could be made that more harm than good has been done in the name of psychology. This is not, of course, to say that scientific enquiry into human nature is a bad thing; it is only to argue that application and misapplication of psychology has done a great deal of damage, that the *'psychologization'* of our contemporary world is a frightening development. It is no great insight that while knowledge itself is value free, it can be used for good or evil: Say the word Hiroshima. Nevertheless, no sane person argues for banning research in physics or stem-cell research, for futilely trying to put the half-eaten apple back on the tree. The difference between abuses associated with physics or biology and psychology is one of insidiousness: The damage 'applied psychology' has done is far more subtle than atomic weapons, but has the potential to be just as destructive to civilization.

Social Engineering.
This is, in my estimation, where the most extensive harm has occurred from psychology's rise to prominence in the last hundred years. The purpose of science is understanding, and a critical test of understanding is prediction—not *a posteriori* explanation. Unfortunately because prediction yields power, it is easy to *conflate* the test with the thing tested—even *inflate* the test to the status of *purpose*. So if the subject of a science is behaviour, and one feels one can predict behaviour, then one assumes one can control behaviour. And so assume the psychologists.

Being able to predict the results of radioactive decay of Uranium in specific conditions gave us nuclear power plants—and the atomic bomb. So as everywhere in science, the power gained by knowledge can be used for good and evil. Psychology, unlike physics, however is not accurately predicting and controlling the behaviour of subatomic particles: it is *inaccurately* predicting and *trying* to control human behaviour. Ethics is important in decisions regarding the application of knowledge in physics, but psychology often goes a step further in presuming ethics is actually part of what it knows and understands—which patently is not true.

That presumption, based on incredible moral *hubris,* is the core of the problem with social engineering. Who do these psychologists, these would-be social engineers, think they are? What evidence can they offer to justify setting themselves up as moral arbitrators qualified to design a society? Except for those empowered politically (dictators, feudal lords, kings, popes, caliphs, and others of such ilk), such *hubris* was once only symptomatic of some few philosophers, who with few exceptions—Lenin comes to mind—did not attempt to impose their moral philosophy and conception of society on anyone.

.

Ironically, psychology's invasion of the public sphere might be even more destructive if its predictive powers were as good as those of the hard sciences, but the reality is that they are not, in most areas, very good at all. Psychologists often explain away all their failures by saying that psychology is still "a young science". They've been saying that since before I was born, and I wish I could say I'm still young. This excuse is getting to be pretty old—and lame. Chart the progress of psychological knowledge next to that of, say, knowledge in physics from the beginning of the twentieth century to now. Hmm. In the year 1900, physics was certainly still a young science as well. There was no relativity theory or quantum theory, and Freud had already published several books. There are still plenty of Freudians out there, but you'd be hard put to find any physicist who still embraces the clockwork view of the universe of 1900. There is no need to list the accomplishments in understanding, prediction and control of the physical universe that have occurred in the last century. What has psychology to offer by way of comparison? That's a rhetorical question. It's no contest.

.

But, nevertheless, psychologists are taken far more seriously than their accomplishments justify. They are the current advisors to the throne. You want to improve your public image and get elected to office? Consult a psychologist. You want to know how to sell a product? Consult a psychologist. You want to know how to do effective propaganda? Consult a psychologist. You want to 'break' a prisoner in interrogation? Consult a psychologist. You want to reshape public attitudes toward some behaviour or group of people? Consult a psychologist. You want to increase the likelihood that a soldier will unquestioningly kill on command? Consult a psychologist. You want to get conformity to some norm. Consult a psychologist who can label any undesired behaviour a form of mental illness. You want to justify censorship with 'scientific' data? Consult a psychologist who can offer 'evidence' of the dangers of free expression. You want to justify just about any legislation that restricts citizen rights? Consult a psychologist: he will find some study that 'links' such freedom with deleterious effects.

The bad science gets mixed with the good. Yes, the advice given occasionally works, but then most of these 'successes' are viewed with horror by those who value human rights and common decency and who are appalled by such amoral manipulation of human behaviour. It is estimated that in World War I, only 20% of the foot soldiers actually fired their guns at the enemy, unable to really try to take another human being's life. Psychologists have helped 'improve' training in boot camp, and that reticence to kill is no longer common among its graduates. *Voila*: My Lai.

To live in a democracy means to live in a dictatorship of the majority.* The only thing that can prevent democracy from being as totalitarian as any conventional dictatorship is the guarantee of human rights, without which majority rule is no more than mob rule. This guarantee is to be found in the American Constitution's Bill of Rights and the United Nations Charter of Human Rights, but the reality is that somebody has been adding small print to this warranty. So as de Tocqueville warned almost two hundred years ago (and Aristotle two thousand years ago) most democracies and their laws are at the mercy of mass opinion—which is something social psychologists really have gained some dangerous insight into controlling.

Such control, as with much of psychology, is primarily based on statistical analysis. The branch of statistics relevant here is inferential statistics: the science of reliably predicting, inferring, the *majority* of outcomes from initial conditions. *Majority* is the key word here. 'Outliers', as they are called, are irrelevant for social engineers, salesmen, and politicians. Majority rules. Democratic elections and governmental policy between elections are now shaped by the data supplied by pollsters, who are nothing more than applied psychologists well trained in inferential statistics. Few think this a good trend for representational democracy.

And how does this current situation affect the creative individual—who is by definition an 'outlier'? It makes him not just an outlier, but an outcast as well. That personality trait called 'disagreeableness' is interpreted as a social evil. I am sure the romantic view of the artist as outsider ('beat' or 'rebel' or 'troublemaker' or 'mentally ill') has

---

* Communism was supposed to be a benign "dictatorship of the proletariat", but of course has historically ended up being just plain old-fashioned dictatorship—and anything but benign.

been reinforced by the culture of conformity that springs from democracy and has been so richly nourished by psychology.
.

Conversion Of Individuality And Creativity Into A 'Problem'.
This is the second general category of harm resulting from the psychologizing of the modern world. It flows out of the acceptance of psychologists engaging in social engineering, for the creative individual does not fit comfortably within a socially engineered society because such a society values homogeneity, not individuality. Group think determines policy. As a corollary to this frightening principle—it should be noted—the value of a work of art, just as the value of a public policy, is to be determined by majority opinion. How does this significantly differ from the value, acceptability, of a work of art being determined by a communist Politburo 'committee'?
.

Since the creative individual is, by definition, deviant from the norm, and since deviance has taken on pejorative connotations (despite it having a simple statistical meaning: the arithmetic difference from the average, the mean), the *exceptionality* of such individuals is viewed with suspicion. Consider the ambivalent response parents have in finding out that the school psychologist has decided their child should be put in a class for "exceptional" children. Should they be concerned or pleased? Is the kid retarded or extremely bright? Thanks to psychology, they are probably worried in either case! Both being bright and being creative have been psychologised into a *problem*: Will little Johnny or Janey be able to fit in?
.

It is interesting that the term 'exceptional' is intended to be 'value-free'—for is not real science value-free? (Cesium isn't better than Helium because it has many more electrons and protons.) But of course everyone knows it is better to be bright than dull! The use of the term 'exceptional', in avoiding this distinction, ends up making a different value judgment: it implies deviance and deviance is not considered a good thing. Statistically, high deviance in a sample decreases one's ability to generalize to a population. Socially, high deviance in a population decreases one's ability to predict and control. So deviance, exceptionality, is inherently a problem to social engineers, even—maybe especially—if that deviance is toward greater intelligence or creativity.
.

Chemists have it easy. Helium atoms are total conformists. In fact they are identical to each other. There aren't any 'slow' ones missing an electron, or 'bright' ones with an extra electron, to mess things up by being reactive when they're supposed to be inert. If people were like atoms, prediction and control would be a piece of cake—and one gets the discomfiting impression that many psychologists really wish folk were more like atoms.

Thus being an outlier in terms of creativity or intelligence is perceived the same way as being mentally ill. It makes no difference whether you believe you are Shakespeare and wrote Hamlet, or you believe you can write a brilliant play just as Shakespeare did. In both cases you are mad—quite literally 'mentally ill'. The psychological community tends to respond to all outliers the same way: by labelling them mentally ill.

The diagnostic Bible of psychiatry (the DSM-IV*) is a crude taxonomy of mental illness. Insofar as it roughly categorizes behavioural complexes, it serves the useful function of improving communication between therapists. But unfortunately it is treated as if it were something much more and is used to label—stigmatize—people, often with deleterious results. It is *not* a periodic table of the elements. The distinction between Helium and Cesium is real, based on sharp, clearly defined criteria. The distinction between Schizotypal and Schizoid 'personality disorder' is *not*! These are descriptive categories of two of the three† so-called "personality disorders" that were chosen by a committee of shrinks to fit in the larger "Odd/Eccentric" category—which is just as artificial and arbitrary.

In pre-psychological society, someone would have been labelled "painfully shy" rather than "schizotypal". And the eccentric would be considered eccentric—not "schizoid". Contrary to what the politically correct would have us believe, it is usually less harmful to dismiss some eccentric as a "nutbar" than to tag him with a diagnostic label such as "schizoid personality disorder". The former label is obviously a personal opinion; the latter implies scientific objectivity.

The tendency to *ex post facto* 'diagnose' the eminently creative with this, that, or the other type of mental illness is a natural extension of this general propensity to 'diagnose' as dysfunctional any deviation from the norm, but this is a topic for a future section on madness and creativity. What is relevant here is the contemporary harm being

---

* The full title is *The Diagnostic and Statistical Manual of Mental Disorders*.
† The other is "Paranoid".

done by this pandemic of psychological diagnoses. Unlike diagnoses in medicine, diagnoses in psychology are not clear cut or usually subject to empirical verification. Doctors may disagree about a diagnosis, but eventually the correct one can be established—even if post mortem. Psychiatrists disagree much more, as studies of the consistency between different diagnoses have repeatedly demonstrated. But, more importantly, rarely is there any clear empirical validation of one labelling over the others.*

.

Furthermore, psychological diagnoses, unlike most medical ones, are never free of value judgments—no matter what your neighbourhood psychologist claims. It is a very different thing to say "Mary is diabetic!" from saying "Mary is neurotic!" The DSM-IV fits the evolutionary model: it is a creature driven by its selfish genes to extend its territory. Every new edition of DSM has significantly increased the number of "mental disorders", offering support for the critics' claim that psychiatrists are simply inventing illnesses. Only rarely, through political pressure, has it dropped a category. A perfect example is that the latest edition no longer labels homosexuality as a psychological disorder, but now labels smoking a psychological disorder. As one British wag, remarked: "Ya youse ta be mentally sick if you *were* a fag. Now ya a nut case if you *smoke* a fag." Only the extremely naïve would think this is real advancement in the science of psychology: it is demonstrably the result of pressure from the homosexual community† and the anti-smoking lobby on the psychiatric committee responsible for determined categories in the DSM-IV.

.

Of course, neither homosexuality nor smoking would be classified as mental illness by any reasonable person—only a committee of shrinks could do that. Nor should almost any of the other recent new 'mental illnesses' added to this diagnostic manual, such as caffeine addiction. I'll admit I may be crazy, but I don't think that such a diagnosis should be based on my craving for a cup of java and a smoke first thing in the morning.

.

The most egregious example of the abuse this type of thinking results in is the practice common in communist countries of labelling dissidents mentally ill and sending them off for 'rehabilitation'.

---

* There are exceptions and these are, as in medicine, the result of a drug treatment failing. An example is manic-depressive disorder, which will respond to lithium therapy, but will not alleviate 'ordinary' clinical depression.
† Three years of intense lobbying before a plebiscite in 1973.

"Comrade, you think your noble leader Stalin is a nasty bit of business? You must be crazy. We'll treat your mental illness."

.

Perhaps the most worrisome section of the DSM-IV is the part that deals with what once was called neuroticism. Neurotics—as the old saying goes—build castles in the sky; psychotics live in them. There is no question that there are real physiological diseases that affect the brain and make their victims completely mad. For example tertiary syphilis, which attacks the brain, did in an appalling number of brilliant people in the 18th and 19th century. But that Schumann, for example, spent his final years in a madhouse really has nothing to do with his musical creativity. And there is no longer any question that schizophrenia is a medical condition, just as is epilepsy. No one would disagree that as medical treatment for such medical problems improves, effort to improve diagnosis is of paramount importance. Let the authors of the DSM-IV work on that. It is the definition of 'neurotic' that is problematic.

.

Virtually all of the eminently creative would be tagged with some kind of personality disorder by an ambitious psychiatrist, DSM-IV firmly grasped in hand. You work hard for incredibly long hours to create books you don't even know will ever be published, and you are a perfectionist—like Flaubert spending a full day getting one sentence right. Then you're clearly obsessive-compulsive, probably schizoid, as well as clearly suffering from Bipolar I. All the coffee you drink and all the tobacco you smoke while working, plus all that alcohol you drink when you finally take a break to socialize and gather material, earns you three more addictive diagnoses. Then when you hit writer's block and you crash, sleeping a lot and moping, we clearly have a case of dysthymic depressive disorder or even Bipolar II. You're a mess, suffering from more mental illnesses than anyone even knew existed in the dark ages before the DSM. That you're Balzac working himself to death is of no importance. With proper clinical treatment you'll behave closer to the norm and forget about that delusionary, grandiose idea of writing *La Comédie Humaine*.

.

Psychologists are a threat to the creative environment. Ecologists consider diversity of species a sign of health in an ecosystem, but it also means the system is chaotic and unpredictable and uncontrollable. What the psychologizing of the world has done is drastically reduce diversity in an attempt to gain more control. The land developer arrives and ploughs under the myriad wildflowers, paves the area and sets up some planters with one species of perennial—and weeds these planters regularly, removing any rogue flowers. It's neat. It's clean. It's profoundly depressing.

.

By trying to 'develop' the social environment, by labelling as weed, and weeding, whatever does not match the plan, too many psychologists are trying to homogenize our world, pave our forests and plant domesticated flowers in cute planters. The good news is that they are usually inept. Weeds are not just called weeds because they are not wanted; they also are annoyingly resilient survivors in the most hostile of environments—much like creative individuals. The bad news is that psychologists on a mission aren't always inept, and they have a lot of public support for their social engineering projects.

USES, SOME

Nothing is pure in this world, and that includes evil. Psychology is not without its redeeming characteristics. The goal of pure science, and psychology, insofar as it manages to be science, is understanding. And we really do understand more about behaviour, including creative behaviour, because of psychological research. I have tried to separate the wheat from the chaff in harvesting this research for inclusion in this book.

There are three areas where this wannabe science has more or less successfully operated as a secret agent for creativity.

- Understanding the nature of the creative person and the creative act
- Understanding the innate mechanisms that induce specific responses to the creative work
- Understanding the personal and evolutionary value of creativity and its products

Understanding the nature of the creative person and the creative act.
This is, of course, also one the major goals of this book, so reference to psychologists' efforts in this area litter this text. Call me a hypocrite. That I don't feel this psychological endeavour has always been successfully prosecuted must be obvious already. However, there are interesting, and sometimes counter-intuitive, discoveries that empirical studies have unearthed.

Understanding the innate mechanisms that induce specific responses to the creative work.
An early psychological school of thought, originated by Wundt and proselytized by Tichener*, is called 'functionalism'. Its primary methodology was introspection, which involved trained subjects 'introspecting': i.e., trying to watch or catch what their minds were thinking and feeling when presented with different stimuli. "Does a green light make you feel calm or agitated?" might be a question asked of an introspecting subject. Much of modern empirical aesthetics derives from this early attempt to link affective response to external stimuli, such subjects judging the relative pleasingness of

---

* Some would say "distorted" by his disciple Tichener.

rectangles with ratios of height to width matching or not with the famous "Golden Ratio" of antiquity.*

All the discoveries made in the fields of sensation and perception have contributed mightily to understanding our responses to art. To cite but one example, the understanding of colour vision (pioneered by Young, Helmholtz and Hering) is of great value to practising artists, although it doesn't seem to have filtered down to the teaching of colour theory in fine arts faculties—another example of the two cultures' lack of communication.

Understanding the personal and evolutionary value of creativity and its products.
What motivates a creator to create? What motivates a percipient to care about the creations? These are profound philosophical questions, and also themes of this book. Psychology actually does proffer some reasonable hypotheses to explain these apparently purposeless motives for our behaviour. Aristotle said "All men by nature desire knowledge." And aesthetic experiences. But to say it is our nature to do so is a very superficial 'explanation', as Aristotle would certainly admit. Why is it our nature? What possible function does creation and appreciation of art or science serve, to us a species, as carriers of our "selfish genes"? No psychological theory adequately answers this question, but psychology does suggest some possible, albeit partial, explanations to this mystery.

It would be negligent to conclude this section without making it clear that psychological research has yielded many other valuable insights into human (and animal) behaviour. The ones briefly mentioned above are just those that seem directly relevant to the theme of this book.

It is at the biological end of the psychological spectrum that the greatest accomplishments are to be found. Neural imaging techniques (such as PET, CAT, and fMRI scans) have yielded real insights into brain function. The discoveries in the field of psychopharmacology over the last five decades have done more for the *really* mentally ill than anything in the rest of history, because—of course—they are *really* physically ill. The deeper understanding of genetics and the machinations of evolution have resulted in the enlightened realization

---

* Two lengths have a 'golden ratio' if the ratio between the sum of these lengths and the longer one is the same as the ratio between the longer one and the shorter. It is approximately 1.62.

that many so-called psychological problems are actually hereditary and medical mishaps, not the result of bad parenting—or weak wills, neuroticism or moral turpitude.*

Research in learning, thinking, language development, and memory has also yielded tangible findings; many with practical applications individuals can use to improve their lives. Of course many of the insights now substantiated empirically really date back to the ancient Greeks and beyond down into the mists of time, but empirical substantiation is still essential to separate belief from fact.

---

* Your child isn't autistic because you are a "cold" parent; it's a brain disorder that is largely genetic in origin. You don't have ulcers because you are a neurotic Type A personality; you just have a bacterial infection in your gut that can be treated with antibiotics.

CASE STUDIES: SIGMUND FREUD & B. F. SKINNER

I feel Sigmund Freud and B. F. Skinner make perfect antagonists, superb examples of two extremes of psychological thinking. Both fancied themselves hard-headed scientists at war with a reactionary Establishment reluctant to face facts. Both were indisputably brilliant and creative. Both stood with one foot firmly planted in philosophy and the other just as firmly planted in what they believed to be real science. Both were artists (interesting writers) as well as scientists. Both were, and still are, usually either revered or despised—and both largely by people who really don't know enough to pass judgement.

But they are polar opposites in their approaches to understanding human behaviour. It is a lovely fantasy to imagine them meeting in a bar and entering into discussion. (I once read a charming memoir about a very nervous Tennessee Williams meeting Ernest Hemingway in a Cuban bar. Swish meets Macho! Apparently and surprisingly, they got along famously. Perhaps Freud and Skinner would have as well.)

Freud put the 'mind'—conscious and especially subconscious—at the centre of his theoretical framework. Skinner banned the 'mind' from any consideration: it was outside the domain of empirical science. Never mind the subconscious; even consciousness wasn't 'scientific'!

Sigmund Freud was born in Moravia in 1877. His family was relatively poor, but like many Ashkenazy Jewish families they put great value on education, and Freud was obviously a bright boy. Unlike many creative innovators, he actually excelled in school from an early age. Despite anti-Semitism and related admission quotas, he was admitted to the University of Vienna at age seventeen. In 1902, after the publication of several of his early books, he returned to this university with a professorship. He had for the previous decade and a half been in private practice, developing his complex theory of human personality on the basis of his interactions with his clients.

There are many excellent biographies of Freud, most notably that by Ernest Jones, and the "Freudian Literature" is huge, for his theories are as influential, if clearly not as scientific or substantiated, as those of Darwin or Einstein. Obviously, there is no room here, nor is it appropriate, to even begin to explicate his many hypotheses about human nature but some background explication of his interpretation of the nature of creativity is necessary.

Freud hypothesized three structures to the human mind: the superego (which can be roughly equated with our conscience); the id (which can be roughly equated with our desire for instant-gratification and our 'animal' nature); and the ego (which is responsible for negotiating a compromise between the demands of the other two). Freud seemed to view all human life to be a constant conflict between the demands of one's id and one's superego, with the meaning of a 'well-adjusted' person being those very few whose egos' could cobble together a working truce. His famous "defense mechanisms" are a list of the lies the negotiator ego uses to help us make it through the day—and night. They include denial, repression, reaction formation, rationalization, projection, displacement, and sublimation.

The last three are particularly relevant to the creative person, as conceived of by Freud. *Projection* refers to the attributing of one's own characteristics to others or other things: the fool thinks everyone else a fool; the sexually obsessed thinks everyone else is obsessed with sex. *Displacement* refers to displacing one's feelings, or expression of those feelings, to a safer object: the man angry at his boss, bites his tongue at work, but comes home and kicks the dog or yells at his spouse. *Sublimation* is really a sub-category of displacement and projection, for it refers to projecting and displacing in a way that is socially valuable: This, according to Herr Freud, is what the artist, the creative individual, does!

You are obsessed with having sex with someone forbidden. You feel great anger toward someone to whom you dare not express your anger. (Freudian sophisticates are thinking Ma and Pa, respectively; but these need not be the only objects of your frustrated desires.) So write a book, paint a picture, or compose a symphony. The world praises you and appreciates you and you have contributed to civilization. But, according to Freud, your real motives remain base and savage, and would embarrass your fans. Well at least until the Freudian critic reveals them in an exposé published in some journal with a Freudian bias. Then your fans are supposed to have a deeper insight into your creative work and empathize with your 'conflicted' soul.

It bears repeating: For a man with a hammer, every problem is a nail. Freud was a clinician and he based his theories on his interaction with emotionally disturbed patients. Just as a NYC cop working a tough neighbourhood is likely to develop a view of human nature that is less than flattering, because of his daily interaction with the dregs of humanity, so too Freud's biased sampling may be why his view of human nature was so dark.

·

You might expect that artists would react negatively to a theory that effectively said their greatest creative accomplishments were just a *symptom* of their mental illnesses. But that was not the case. A large portion of the creative community embraced Freudian theory enthusiastically. Before the academic 'credentialization' of psychoanalysts*, numerous artists even became 'lay analysts'. Even such unlikely characters as Henry Miller and Anais Nin dabbled in psychoanalytic therapy. Also artists, especially in the first part of the twentieth century, seem attracted to the dark mythos and pessimistic view of humanity that Freudian theory had as its cornerstone. This was at a time when expressionism, the idea that art was the artist's expression of his profound suffering, a suffering resulting from having greater sensitivity than most other mortals. Artists liked the conception of themselves as heroes struggling against evil dragons— but now not real-world evil, but rather the evil within us all. Many artists seemed to find this image of themselves as tormented, overly sensitive beings perversely flattering. Many still do.

·

Creative scientists were less impressed. For one thing, Freud, and his disciples, really had very little to say about them. (Ignore me; I'll ignore you.) The scientist may have perceived himself as a warrior, but his battles were being fought in the real, empirical world—not in the dark cellars of his soul.

·

It was inevitable that someone from the non-clinical stream feeding the new torrential river of psychology would surface. Psychology as biology and 'hard' science had as long—or even longer—a history as clinical psychology. Researchers such as Weber, Helmholz, Fechner, and Pavlov tried to understand human—and non-human— behaviour in more concrete, objective terms. Freud was a wannabe scientist, but was essentially, at heart, an artist dealing in metaphor much as a poet would. Others interested in this new would-be

---

* Now "psychoanalysts" have to be psychiatrists, and to be a psychiatrist one has to have a medical degree!

discipline chose to emulate the 'hard' sciences and shied away from such unobservable hypothetical constructs as id or libido or the subconscious.

Psychology suffers from a serious inferiority complex, despite its appallingly great influence on our lives. The general public takes psychology very seriously, certainly much more than is sensible or healthy, but psychologists still feel like the new kids on the block who have to prove their mettle with the big kids. That most psychologists so pathetically look up to 'real' scientists (the physicists, astronomers, chemists, biologists, etc.) is evidenced by their obsession with statistics, jargon, and the minutiae of format in published articles.* Meanwhile most of those they look up to as 'real' scientists refuse to take psychology seriously.† It must be *very* annoying.

Behaviourism, which dominated academic psychology for decades, is an extreme example of this determination to be really 'scientific'. Behaviour is what psychology is about, so say they—and actually it is a reasonable thing to say. As wannabe physicists and chemists they refused to even consider anything other than the sensory input and behaviour output. Hell, you can't get inside an organism's mind, an impenetrable 'black box', to observe anything!

Of course this is absurd in terms of deep scientific explanation. That one can predict B from state of A is the most superficial type of explanation. Knowing that knocking an apple off a tree will cause it to fall to the ground may be useful, and could be considered

---

* The standard formatting Bible for publication is the huge *American Psychological Association Style Manual*, which is 'revised' with great frequency (presumably to fund the APA) regarding totally arbitrary and persnickety rules about what is, this year as opposed to last, to be set in bold-face or italics and other fundamentally trivial formatting issues.

† "It's a great game to look at the past, at an unscientific era, look at something there, and say have we got the same thing now, and where is it? So I would like to amuse myself with this game. First, we take witch doctors. The witch doctor says he knows how to cure. There are spirits inside which are trying to get out. ... Put a snakeskin on and take quinine from the bark of a tree. The quinine works. He doesn't know he's got the wrong theory of what happens. If I'm in the tribe and I'm sick, I go to the witch doctor. He knows more about it than anyone else. But I keep trying to tell him he doesn't know what he's doing and that someday when people investigate the thing freely and get free of all his complicated ideas they'll learn much better ways of doing it. Who are the witch doctors? Psychoanalysts and psychiatrists, of course." From *The Meaning of It All: Thoughts of a Citizen Scientist* by the great physicist Richard Feynman.

scientific knowledge, but it hardly constitutes a deep understanding of the 'black box' of gravity. Psychology's *raison d'etre*, if it has any real universal justification for existence, it is the understanding the 'black-box' of the 'O' (the organism) in the S-O-R sequence.

.

Behaviourists had a fear (phobia?) of saying anything about anything that could not be readily observed directly—and that emphatically included the 'mind'. The fact is that no one has seen a proton, yet physicists don't shy away from describing it and explaining it in terms of even less observable critters. Billions of dollars have been spent to build the Large Hadron Collider, a 27 circular kilometre tunnel, which will fling these never-directly-observed protons at each other at near light speed so to observe indirectly the effects of them colliding. Things we can't observe directly can have real effects that allow us to infer a lot about their nature. Apparently this must never have been pointed out to the behaviourists, so determined to be 'scientific'.

.

Now B.F. Skinner certainly was no fool, nor any more naïve than Freud regarding the nature of scientific investigation, but he, for whatever reason, naturally gravitated toward the simplistic, reductionist behaviourist approach of less-gifted people such as John B. Watson, the fellow who is usually credited with initiating this school of psychology.

.

Burrhus Frederic* Skinner was born in rural Pennsylvania in 1904. He took a B.A. in English literature at Hamilton College in New York and promptly moved to Greenwich Village to pursue the vocation of writer. Like Freud, he had literary aspirations that eventually were channelled into the relatively new discipline called 'psychology'. While living in the Village he read Bertrand Russell and John B. Watson, a somewhat strange mix that seems to have inclined him to try to understand behaviour in scientific and philosophical terms, rather than literary ones. He returned to school, receiving at the age of 27 a Ph.D. from Harvard, the institution where he subsequently spent the majority of his academic career.

.

Skinner's influence on psychology cannot be underestimated; and although it could be argued that his wide-spread influence did much harm, there is no denying he greatly increased our understanding of many aspects of behaviour. This is not the place to outline his solid

---

* One can understand why he preferred to just use his initials.

scientific contribution to understanding the type of learning called operant conditioning. The Russian physiologist Pavlov had previously developed a solid theoretical framework for what is now called classical conditioning, based on the linking of learned stimuli to innate responses, but this theory did not take into consideration the role of reward and punishment in conditioning. Skinner's operant conditioning theory did, and did so with great predictive validity and practical application. Applications of his insights have changed, for the better, and informed approaches to everything from dog training through child rearing practices to new psychotherapeutic methods for dealing with phobias. But what is relevant here is his own attitude toward creativity and the relationship of his insights to understanding the creative person.

Freud could be called a 'soft' determinist. He clearly believed that we all were shaped by early environmental influences, with the first five years of our lives probably being the most important. He also believed in certain inexorable laws of nature, *human* nature, which keep us in psychological chains. But he also believed these chains could be—if not broken—loosened, for that was the purpose of psychoanalysis. Skinner, on the other hand, was a 'hard' determinist. He believed that all was predetermined, that the idea of free will was contrary to scientific principles. But of course, like all such determinists, he couldn't 'walk the talk'. In fact, his reputation rests more on his works promoting *willing* use of 'social engineering' to fix what is wrong with the world—something which of course is impossible if the way the world is, and will be, is inexorably set from the beginning of time.

Skinner is known to the general public not so much for his important scientific research on operant conditioning as for his popular and controversial books: *Beyond Freedom and Dignity* and *Walden Two*. The former, with its deliberately provocative title, outlines his philosophy. The second, with a title alluding to Thoreau's famous memoir about living outside of conventional society, is a utopian novel based on his ideas. Skinner maintained that free will and human dignity were unscientific concepts, and that only by applying solid scientific principles (primarily his own operant conditioning principles) could we arrive at a humane and decent society, one free of war and social strife, one which combined minimal consumption of resources, communal sharing of labour and the products of that labour, total egalitarianism, healthy social and family relationships, and meaningful work and leisure. Several attempts were made to build utopian communities based on this book, most of which—not surprisingly—failed, but one, the "Twin Oaks" commune (population a mighty 85 and located in Virginia), has survived since its founding in 1967.

.

There is no question that B.F. Skinner was one of the 'good guys' in terms of his moral values and his sincere desire to make the world a better place by the application of solid scientific principles. Those for whom his name is anathema are those who blame him for all the very real evils of social engineering, but this is a bit like blaming Jesus for the Inquisition. Skinner's philosophy may have been naïve, but his science was sound, if limited in scope, and his own suggestions for applying this science fundamentally benign.

.

His view of the creative individual was that of a person freed of the bonds of bad early conditioning and with the leisure and energy to live a full life and contribute to civilization. The biographical evidence is that his own life was much like how he envisioned the creative individual's life should be. Absurd rumours that his application of his principles to his own children led to very dysfunctional adults still circulate. One such rumour has his daughter Deborah committing suicide as a result of being raised in a "Skinner Box" (a human size version of the now standard apparatus for experimenting on rats learning in operant conditioning experiments). Skinner did in fact design an environmental crib that gave the infant more room and more comfort than conventional cribs. (But it certainly didn't have a shock grid floor, as do most rat Skinner boxes, or levers that had to be pressed to get food pellets.) Both his daughters are alive and well and successful—and have publicly expressed love and admiration for their father.

.

Skinner and Freud are examples of *scientifically* creative individuals who have greatly influenced the conception of the creative *artist*, largely because of their own experience of creativity. Both had literary ambitions, and both had talent in that domain which their best writing demonstrates. But both chose the scientific over the artistic path.

.

Freud's actual influence was greater among artists because his dark, mythic view of the world was more congruent with the self-image artists had at the time—and which many still do. Skinner was a better scientist, but his view of creativity is embraced not by artists so much as by educationists and those who would like to believe creativity is universal—and an unreservedly positive thing. Neither perspective is accurate, for both are simplistic; but they are like two pure and intense colours at the extreme opposite ends of the spectrum of opinion regarding creativity. The next section, the central pane of this book, will attempt to unweave this spectrum and display the whole rainbow, which is both bright and dark, but always illuminating.

EPILOGUE: THE PLAY'S THE THING; TO PLAY IS THE THING

"The play's the thing / Wherein I'll catch the conscience of the King".
—William Shakespeare (*Hamlet*)

"The true object of all human life is play."
—C. K. Chesterton (*All Things Considered*)

The king has no conscience and will never be caught, but play on we will, *for it is our very nature to do so*. The creative urge is an extension of the procreative urge. Darwin had it right: the two forces that "through the green fuse drives the flower"* of our evolution and our creativity are our own survival and the survival of our creations, be they living, breathing offspring or those more abstract offspring—our artistic creations and scientific discoveries.

Both are the result of deadly serious play. The playfulness of the young of every species is necessary practice of the skills required for subsequent survival and procreation. Human beings, especially creative human beings, never give up this playfulness. Life is a game. The stakes may be high, but it is a game nonetheless.

Just as love is a very serious game, as anyone who has ever been in love knows, so too are scientific and artistic endeavours. For the creative person, creativity is at least as much a driving force as procreativity—or even individual survival.

The rules of the game are different in art than in science, and even different in the different arts and sciences, and these differences are what I've tried to explore in this book. But I hope I have also shown the common grounding of both in playfulness; playfulness has the common goal of winning an apprehension of the world.

Whatever random genetic mutations are responsible for the extension in us of the neonatal propensity for play into human

---

* See Dylan Thomas' poem of that name.

adulthood is not now known. But it certainly has worked in terms of extending the influence and success of our species. It is ironic, of course, that our failure to 'grow up' is what has led to our dominion over the earth. And, of course, it will be even more ironic, albeit more bitterly and tragically ironic, if at the final curtain call, our creative playfulness results in the end of not only our survival as a species, but that of all life on our planet *and* all the marvellous and miraculous things we have created.

Maybe one can only sigh and say that what will be, will be: "Que sera, sera!" Who knows what lies at the very bottom of Pandora's Box or what poison the last bite down near the core of the Apple of Knowledge contains? Nevertheless, we, as human beings have committed. There in no turning back. We will play on till the end of time, or the end of *our* time. The world's a stage, and although every single creative production hasn't been something to add to the asset column of the moral ledger, I, for one, hope the theatre doesn't go bankrupt and get closed down—at least not until the universe itself finally collapses into an entropic, boring heat-death.

So on with the show!

AFTERWORDS

In other words, stuff stuffed at the end.

ACKNOWLEDGEMENTS

I think virtually everyone, upon beginning to read a book, skips the acknowledgements—everyone, that is, except those who feel they should be acknowledged. Better, it seems to me, is to do as they do in the movies: put the credits at the end.

I'm limiting my acknowledgements so those who have been so particularly helpful and supportive that they shouldn't be buried in a huge list that no one will read. However, I also would like to extend an all-purpose, less specific, thank you to some other folk as well.

There is no way I adequately thank my wife, Ursula, for all she has done for me. She has always been an indefatigable editor. Without her, I would've been continuously embarrassing myself in print from the very start. But most importantly, she has been the inspiration for all my work. She redeems my life. She is my Beatrice—and I got to actually marry her and not just admire her from a distance.

I also owe a debt of gratitude to our two 'kids', Christiaan and Katherine. They certainly deserve credit for their tolerance of their eccentric father, but, more importantly, for their passion for all the aesthetic and intellectual, and even purely physical pleasures, life has to offer. They have been an inspiration to me.

My family has taught me that loving another and nurturing that 'selfish gene' is the most creative thing we can do. And the most rewarding.

I also am incredibly grateful to the many scientists and artists whose creations were the ultimate inspiration for this book—and even for my way of life.

Thanks, too, to those several teachers I had long ago who through their own passion for art and science instilled in me the respect I have for creative endeavour, as well as to my students who have had to listen to me test out my ideas on them in what I'm sure were sometimes rather rambling lectures in my Psychology of Art class.

I should also acknowledge Nipissing University for giving this writer a very rewarding day job for 40 years. And much of the first draft of *Secret Agents* was written during a sabbatical leave.

SELECTED ANNOTATED BIBLIOGRAPHY

In my judgement the following works are well worth reading (or at least sampling) and *not* because I think they offer supporting evidence for my various theses. (Some do not.) I recommend them because I believe they are all entertaining and enlightening works that are at least tangentially related to the content of this book. Some are about creativity and creative individuals and some are the actual works of the creative individuals mentioned in *Agents*.

Recommending books is way too much fun, for one loves to plug books and writers one admires, but I've tried not to over-indulge in this pleasure. Life is very short—even if one reads quickly—and so I've grouped my recommendations by the sections in *Agents*, so that if something along the reader's way triggers a particular interest, he or she can easily follow up on what inspired me. I've almost always limited my recommendations to a maximum of six per chapter, usually with two of these referring to the subjects of the 'case studies' for that chapter. The list is still long, but rest assured—there will *not* be a quiz!

I admit that what I have chosen to include is extremely arbitrary and idiosyncratic and reflects my own reading rather than some omniscient overview of the many topics touched on in *Agents*. Anyone intimate with a particular chapter's theme could surely cite more relevant and better works, even though I've sincerely tried to select, given my own limited knowledge, excellent works I believe touch most of the bases of each chapter. I've included a very brief justification for the inclusion of each work in this list.

I've usually only given the author and title of the work, for in The Amazonian Internet Age that is sufficient. The only exceptions to this are when more information is really required to easily find the recommended text.

- **The Parting Of The Waters**
  - General suggestions and remarks.
    - o I have chosen water as my primary metaphor, and there is no dearth of artistic works that do the same. From a scientific standpoint, water is both a very common compound on earth, but also an unusual one in many of its properties, and so it has been studied extensively. There is even a seven-volume compendium on the nature of water (Franks, F (Ed), *Water, A comprehensive treatise*, Plenum Press, New York, 1972-1982).
  - Recommended reading.
    - o Snow, C. P. *The Two Cultures and a Second Look, an Expanded Version of the Two Cultures and the Scientific Revolution* – This is the famous Rede Lecture that addressed the unfortunate divorce of science and art in modern times and should be required reading for the many who misinterpreted it as an attack on whichever creative domain they happen to work in.

The Naming Of Parts
- General suggestions and remarks.
  - o Too much is relevant to the topics dealt with in this chapter, for me to do more than present a very idiosyncratic sampling of recommendations. Regarding the case study 'antagonists' representing art and science, it is only reasonable to suggest the reader revisit the poetry of Keats—not that his poetry isn't justification in and of itself. It is not, however, reasonable to expect the contemporary reader to take on Newton's masterpiece, *Philosophiæ Naturalis Principia Mathematica*, so instead I have included a respected biography of Newton in the list below.
- Recommended reading.
  - o Beveridge, William I. B. *The Art of Scientific Investigation* – This little book published back in 1957 remains one of the best and most straightforward descriptions of the scientific endeavour I've come across. It clarifies the common terms and keywords of scientific methodology, without ever getting philosophical or even technical, and is full of

exemplary anecdotes that humanize the scientific project.

- o Giddings, Robert (Editor). *The Letters of John Keats: A Selection* – There is nothing like the personal correspondence of an artist to get a glimpse into his—to use Keat's own phrase—"teeming brain".
- o Gleick, James. *Isaac Newton* – Gleick is one of best science writers around today. He knows his science, and he knows how to write biography.
- o Motion, Andrew. *Keats* – This is a widely acclaimed and extensive biography of the great poet.
- o Sternberg, R.J. *The Nature of Creativity* – Robert Sternberg is one of the most eminent psychologists to tackle the difficult problem of defining creativity in anything remotely resembling scientific terms. He has also taken on the task of redefining intelligence—the subject of the next chapter, where I recommend another of his books.

- **Of Two Minds About Split Brains: IQs And CQs**
- General suggestions and remarks.
    - o There are plenty of books dealing with both intelligence and creativity worth reading. I've only included ones here that specifically address the question of *defining* these ambiguous concepts.
- Recommended reading.
    - o Csikszentmihalyi, Mihaly *Creativity: Flow and the Psychology of Discovery and Invention* – Mister C. is a big name in the field of creativity, partially because he combines some real science with what many readers can interpret as a guide to creativity. He makes the valid point that a significant creative product requires three components: a culture ready for it; an individual capable of creating it; and colleagues working in the same domain to evaluate and appreciate it. Unfortunately in describing the creative personality he takes the can't-miss shotgun approach by 'describing' the creative person as being at both ends of any personality dimension he selects: sometimes introverted, sometimes extroverted; sometimes confident, sometimes insecure, etc.
    - o Da Vinci, Leonardo (translated by Jean Paul Richter). *The Notebooks of Leonardo da Vinci* – These

fascinating documents are in the public domain and can be downloaded from Project Gutenberg.

- o Galton, Francis. *Hereditary Genius* – This classic is still worth reading. It has influenced much of subsequent thinking about exceptional creativity and the whole idea of 'genius', as well as the nature/nurture debate.
- o Gardner, Howard. *Multiple Intelligences: New Horizons* – This book presents the latest version of Gardner's worthwhile division of intelligence into component parts, of which only three of the nine he proposes can conventional IQ tests even purport to measure.
- o Gazzaniga, M.S. (1967). The split brain in man. *Scientific American*, 217, 24-29. Available as a *Scientific American Offprint*. – Presented here is the real science behind 'split brains'.
- o Sternberg, R. J. *Beyond IQ: A Triarchic Theory Of Human Intelligence* – This is Sternberg's attempt to break down the concept of intelligence into three broad component parts (ones which better match what we mean when we call someone intelligent): 1) analytical intelligence (which is closest to what IQ tests purport to measure); 2) creative intelligence; and 3) practical intelligence.

- **The Split (In) Personality**
  - General suggestions and remarks.
    - o A number of books listed in other sections of this bibliography also deal with the themes in this chapter of *Agents,* so I've only included works by the subjects of the two case studies and a book of 'case studies' by a major theorist on creativity. For anyone interested in the topic of personality testing, there are numerous textbooks that deal with the topic, but a quick overview can be found in a chapter in almost any Introductory Psychology text—although it will be more respectful of the field than I am.
  - Recommended reading.
    - o Gardner, Howard. Creating Minds: An Anatomy of Creativity Seen Through the Lives of Freud, Einstein, Picasso, Stravinsky, Eliot, Graham, and Gandhi – What is refreshing about this book is Gardner's deliberate selection of outstanding

individuals from diverse creative domains. Another positive feature of this tome is his willingness to deal with the more unsavoury characteristics of his subjects. Unfortunately sometimes the psychological analyses he offers along with the biographical information sound too much like facile psychologizing for my taste.

- o Rank, Otto (with Charles Atkinson). *Art and Artist: Creative Urge and Personality Development* – This work offers some insight not only into the artistic personality but also into the strange relationship between psychodynamic theory and artists at the beginning of the twentieth century.
- o Rexroth, Kenneth. *Bird in the Bush: Obvious Essays and Assays* – These two collections of Rexroth's essays are worth a thousand academic essays: they are as witty as they are insightful. Additionally recommended are his *Classics Revisited* and *More Classics Revisited*, both worth more than a degree in World Literature because of the insights they offer into literary masterpieces both inside and outside the Western literary canon. Finally, *The Complete Poems of Kenneth Rexroth* (edited by Sam Hamill and Bradford Morrow) reveals an intellectual with a poetic sensibility just as sharp as his critical one.

- **The Solemn Frivolity Of Art And Charming Frigidity Of Science**
  - General suggestions and remarks.
    - o I've selected for this list one huge book on scientists, and three books about the subjects of the chapter's case studies. (This should be sufficient summertime reading for all but *summa cum laude* graduates of Evelyn Wood's Speed Reading School. I should add that all of these books are great for just dipping into and browsing.) What would have balanced this selection would've been a big survey book on 'artists' (including writers, composers, etc.) emphasizing their lives. Of course no such book exists, for it would be more than merely 'big'. So for a closer look at the persons behind the art they create, I can only recommend sampling biographies, autobiographies or collections of interviews. (For

example, *Writers at Work: The Paris Review Interviews* series of books is excellent.)

- Recommended reading.
    - o Gribbin, John. *The Scientists: A History of Science Told Through the Lives of Its Greatest Inventors* – The author is probably best known for *In Search of Schrodinger's Cat*, a wonderful and understandable exploration of the weird world of quantum physics, but he is the author of a hundred other books on science, so to call him prolific is a gross understatement. But it would be very wrong to conclude from this phenomenal output that he is a hack or to assume that the 672 pages of *The Scientists* is sloppy or tedious.
    - o Cage, John. *Silences: Lectures and Writings* and *A Year From Monday* – These two charmingly eccentric books are first-rate literature and will probably be remembered longer than his actual musical compositions.
    - o Feynman, Richard P. (with Ralph Leighton, Edward Hutchings, Albert R. Hibbs). *Surely You're Joking, Mr. Feynman! (Adventures of a Curious Character)* – This is a sort of autobiography of Feynman filled with amusing anecdotes that should eradicate any misconception of the scientist as some humourless, tight-assed guy in a stained lab coat.
    - o Gleick, James. *Genius: The Life and Science of Richard Feynman* – Gleick is a fine writer and here gives a very entertaining biography of one of the great physicists of our time—and one that illustrates many of the points made in *Agents* about the character of the creative individual and the playfulness of scientists.

- **Common Ground At The Confluence Of Art And Science**
    - General suggestions and remarks.
        - o Where to begin?! The writings of many scientists should qualify as Literature with that capital 'L'. And within the canon of great literature many writers have often taken as their theme the natural world, demonstrating observational skills worthy of any professional naturalist. Then there is science fiction, a genre that deliberately aims to merge literature and

science—and often philosophy. The plethora of possible inclusions for this chapter's list is support for my central thesis of the intertwining of art and science. It also means I don't know where to begin—and that inclusion in this painfully tiny list means nothing more than I'm taking this opportunity to plug a few books I love. (I have, of course, included two books by the subjects of the chapter's case studies.)

- Recommended reading.
  - o Clarke, Arthur C. *Childhood's End* – This is one of his science fiction novels for which I have an especial fondness and one that deals less with science and more with philosophical issues. Of course *2001: A Space Odyssey*, the film that he collaborated on with Stanley Kubrick, is considered by film critics to be a modern cinematic masterpiece. (It should be noted that the book was completed after the film was released and varies on some key points.)
  - o Dillard Annie, *Pilgrim At Tinker Creek* – Following in the literary footsteps of Thoreau, but with prose more subtle and resonant, this great contemporary essayist observes deeply and meditates deeply on the natural world in a style that can only be described with that archaic word 'sublime'.
  - o Eiseley, Loren. *The Immense Journey: An Imaginative Naturalist Explores the Mysteries of Man and Nature* and *The Night Country: Reflections of a Bone-Hunting Man* – The former is his best known book, a modern classic which has sold millions of copies. The latter contains what I consider some of the most beautiful and touching personal essays written in the twentieth century.
  - o Hoagland, Edward. *Hoagland on Nature: Essays* – Here at last is a gathering together of some of the best nature essays by a writer who seems to scatter his gems far and wide in periodicals.
  - o Lightman, Alan. *Einstein's Dreams* – Alan Lightman is a physicist, novelist and essayist who has made major creative contributions in all three domains. This literary novel, translated into 30 languages, is as unlikely a best-seller as one could imagine given its premise and structure, but its success is

understandable once one has read a few pages. Its popularity, I might add, supports my contention that there is a growing audience for creative works at the confluence of art and science.

- o Stapledon, Olaf. *Last and First Men* and *Star Maker and Sirius* – These three novels are old favourites of mine—and neglected science-fiction classics from the thirties and forties. Stapledon was a philosopher and these seminal works have inspired some of the best serious and literary science-fiction writers that followed him. They have even inspired scientists with scientific ideas—an example of art affecting science even at the most abstract level. For example, the physicist and mathematician, Freeman Dyson, credits his concept of "Dyson Spheres" to reading *Star Maker*. Genetic engineering and terraforming are ideas presented well ahead of their time in *Last and First men*. And anyone who loves dogs, as I do, will love *Sirius*: a waggish tail (sic) of a dog whose intelligence is raised to at least human levels.
- o Vonnegut, Kurt. *Cat's Cradle* – Vonnegut wrote this short, disturbing novel early in his career, but many people, myself included, still consider it his best book. Besides, every literate person should know about the novel's polymorph of water, "ice-nine"— a terrifying metaphor for so many things our quest for knowledge has wrought.

- **Quicksand On The Shore: The Social Evils Of Social Science**
  - General suggestions and remarks.
    - o The writers most likely to see through the posturing and pretensions of psychologists are, not surprisingly, the great satirists and cynics. Jonathan Swift does a nice job on academics and those who will come to be called 'scientists' in the "A Voyage to Laputa, Balnibarbi, Glubbdubdrib, Luggnagg and Japan" section of *Gulliver's Travels*. H.L. Mencken is always worth reading for his irreverence toward faddish would-be science. Mark Twain also takes some nice pot shots; his brief essay response in *Answers To Correspondents* to the "Moral Statistician" is charming. Then there is the dystopian genre,

where the potential for extreme totalitarianism from the application of 'psychological' manipulation is explored. The classic example is, of course, George Orwell's *1984*. Unfortunately psychology is now such a sacred cow that I know of no single contemporary non-fiction book devoted to an extended and serious criticism of its elevated status and the harm that has resulted from this power.

- Recommended reading.
  - o Freud, Sigmund. *Writings on Art and Literature* (Stanford University Press, 1997) – Freud was very prolific, and his writings dealing specifically with art and artists are scattered throughout his huge output. This collection brings together under one cover Freud's major essays on this theme.
  - o Glover, Nicola. "Chapter 1: Freud's Theory of Art and Creativity" in *Psychoanalytic Aesthetics: The British School* – This nice summary of Freud's view of art and artist is only available at time of writing online at— *http://www.human-nature.com/free-associations/glover/chap1.html*
  - o Jones, Ernest. *The Life and Work of Sigmund Freud* – This is the classic, definitive biography of Freud. Be warned: Jones is a believer.
  - o Skinner, B.F. *Beyond Freedom and Dignity* and *Walden Two* – The former book lays out Skinner's philosophy and the latter is his utopian novel based on this philosophy. Both are very readable and reveal Skinner to be a sheepish humanist in wolf's clothing.

AUTHOR'S NOTE

"I believe that literature, like science, is a way of exploring different perspectives; and I believe that the results of these literary explorations, like the results of science, are always inherently tentative. It is for this reason that I choose to call my major works 'hypotheses'. *Secret Agents Past: The Parting Of The Waters* is Hypothesis 15."

ABOUT THE AUTHOR

Ken Stange is the author of 15 previous books of poetry, fiction, and non-fiction, as well as hundreds of publications in literary and scientific journals. He was the winner of the 2011 Exile/Vanderbilt prize for short fiction, and he is also a visual artist and Professor Emeritus at Nipissing University where he continues to teach "The Psychology of Art" as an online course. His special interest is the relationship of art and science and creativity.